EARTH
THEN AND NOW

MITCHELL BEAZLEY

EARTH
THEN AND NOW
Potent visual evidence of our changing world

Text: Fred Pearce
Foreword: Zac Goldsmith

Downtown Dubai This scene was desert until the mid-1990s.

Contents

Map Contents 6

Foreword by Zac Goldsmith 8

Introduction 10

Environmental Change 20

EARTH THEN AND NOW

Fred Pearce

First published in Great Britain in 2007 by Mitchell Beazley,
an imprint of Octopus Publishing Group Ltd,
2–4 Heron Quays, Docklands, London E14 4JP

Copyright © Octopus Publishing Group Ltd 2007

A CIP catalogue record for this book is available from the British Library.

ISBN-13: 978 1 84533 246 4
ISBN-10: 1 84533 246 6

Commissioning Editor **Jon Asbury**
Executive Art Editor **Yasia Williams-Leedham**
Senior Editor **Suzanne Arnold**
Designer **Colin Goody**
Editor **Andrew Puddifoot**
Proofreader **Emily Anderson**
Picture Research **Giulia Hetherington, Elaine Willis, Jenny Faithfull**
Indexer **Alan Thatcher**
Production **Peter Hunt**

Set in Myriad Pro

Colour reproduction by Sang Choy, Singapore
Printed and bound by Toppan, China

Forces of Nature 162

Urbanization 66

Land Transformation 118

War and Conflict 208

Leisure and Culture 246

Index and acknowledgments 282

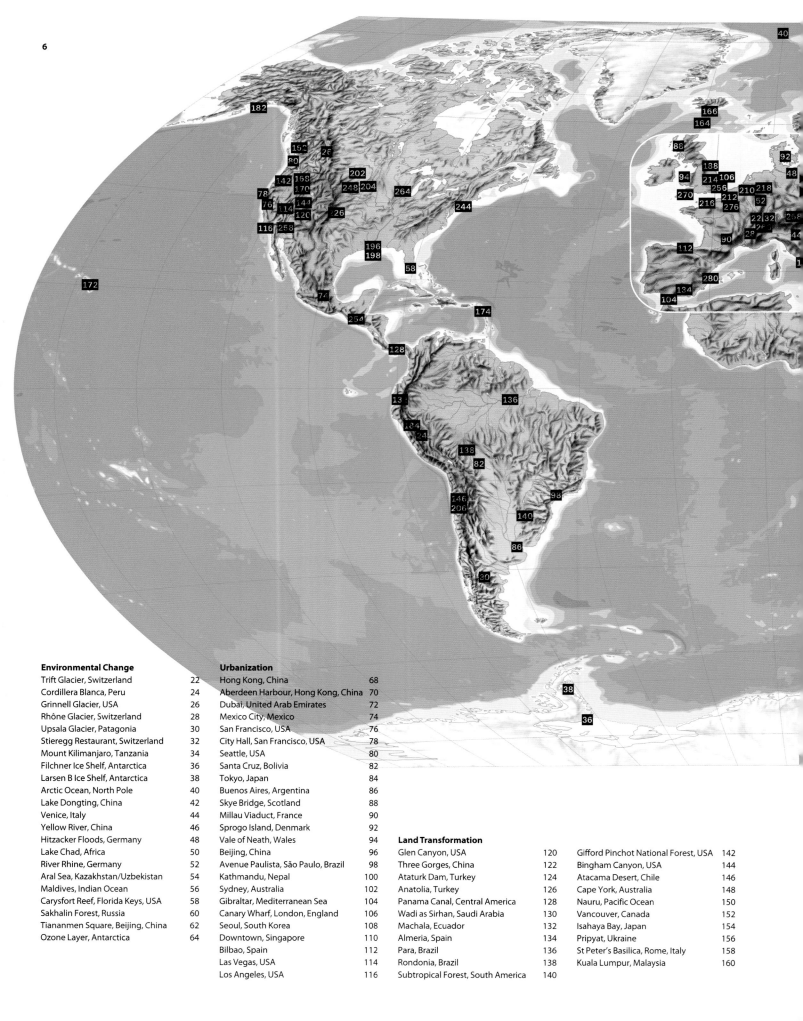

Environmental Change

Trift Glacier, Switzerland 22
Cordillera Blanca, Peru 24
Grinnell Glacier, USA 26
Rhône Glacier, Switzerland 28
Upsala Glacier, Patagonia 30
Stieregg Restaurant, Switzerland 32
Mount Kilimanjaro, Tanzania 34
Filchner Ice Shelf, Antarctica 36
Larsen B Ice Shelf, Antarctica 38
Arctic Ocean, North Pole 40
Lake Dongting, China 42
Venice, Italy 44
Yellow River, China 46
Hitzacker Floods, Germany 48
Lake Chad, Africa 50
River Rhine, Germany 52
Aral Sea, Kazakhstan/Uzbekistan 54
Maldives, Indian Ocean 56
Carysfort Reef, Florida Keys, USA 58
Sakhalin Forest, Russia 60
Tiananmen Square, Beijing, China 62
Ozone Layer, Antarctica 64

Urbanization

Hong Kong, China 68
Aberdeen Harbour, Hong Kong, China 70
Dubai, United Arab Emirates 72
Mexico City, Mexico 74
San Francisco, USA 76
City Hall, San Francisco, USA 78
Seattle, USA 80
Santa Cruz, Bolivia 82
Tokyo, Japan 84
Buenos Aires, Argentina 86
Skye Bridge, Scotland 88
Millau Viaduct, France 90
Sprogo Island, Denmark 92
Vale of Neath, Wales 94
Beijing, China 96
Avenue Paulista, São Paulo, Brazil 98
Kathmandu, Nepal 100
Sydney, Australia 102
Gibraltar, Mediterranean Sea 104
Canary Wharf, London, England 106
Seoul, South Korea 108
Downtown, Singapore 110
Bilbao, Spain 112
Las Vegas, USA 114
Los Angeles, USA 116

Land Transformation

Glen Canyon, USA 120
Three Gorges, China 122
Ataturk Dam, Turkey 124
Anatolia, Turkey 126
Panama Canal, Central America 128
Wadi as Sirhan, Saudi Arabia 130
Machala, Ecuador 132
Almeria, Spain 134
Para, Brazil 136
Rondonia, Brazil 138
Subtropical Forest, South America 140

Gifford Pinchot National Forest, USA 142
Bingham Canyon, USA 144
Atacama Desert, Chile 146
Cape York, Australia 148
Nauru, Pacific Ocean 150
Vancouver, Canada 152
Isahaya Bay, Japan 154
Pripyat, Ukraine 156
St Peter's Basilica, Rome, Italy 158
Kuala Lumpur, Malaysia 160

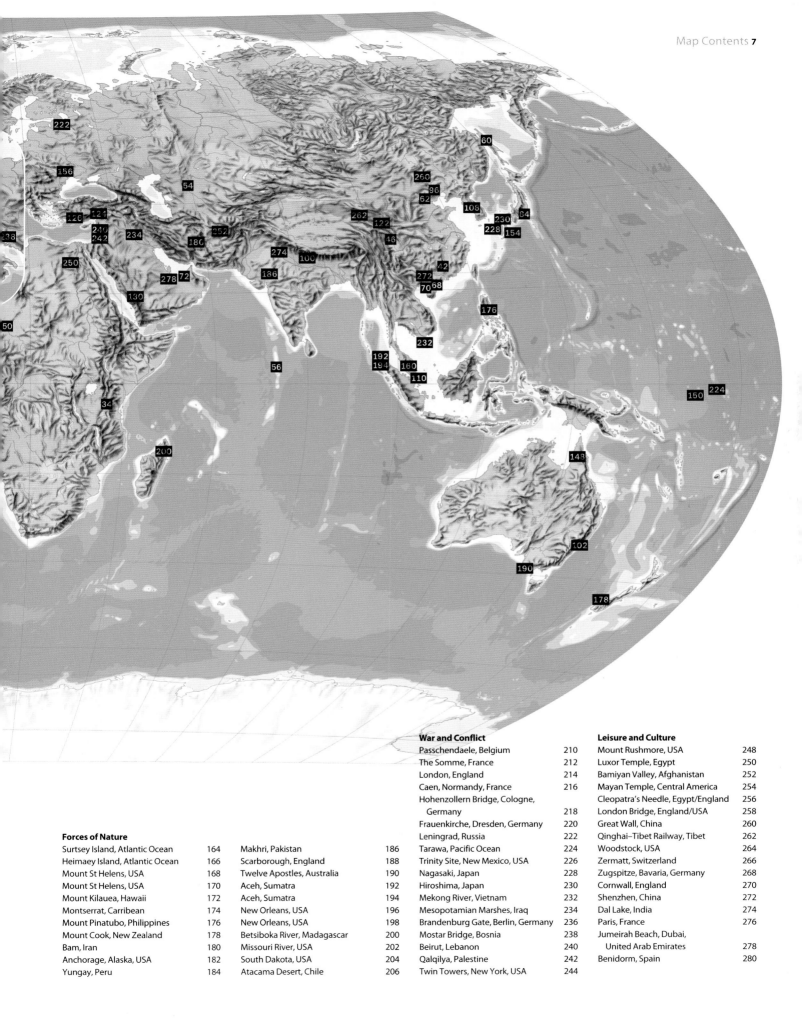

Forces of Nature

Surtsey Island, Atlantic Ocean	164	Makhri, Pakistan	186
Heimaey Island, Atlantic Ocean	166	Scarborough, England	188
Mount St Helens, USA	168	Twelve Apostles, Australia	190
Mount St Helens, USA	170	Aceh, Sumatra	192
Mount Kilauea, Hawaii	172	Aceh, Sumatra	194
Montserrat, Carribean	174	New Orleans, USA	196
Mount Pinatubo, Philippines	176	New Orleans, USA	198
Mount Cook, New Zealand	178	Betsiboka River, Madagascar	200
Bam, Iran	180	Missouri River, USA	202
Anchorage, Alaska, USA	182	South Dakota, USA	204
Yungay, Peru	184	Atacama Desert, Chile	206

War and Conflict

Passchendaele, Belgium	210
The Somme, France	212
London, England	214
Caen, Normandy, France	216
Hohenzollern Bridge, Cologne, Germany	218
Frauenkirche, Dresden, Germany	220
Leningrad, Russia	222
Tarawa, Pacific Ocean	224
Trinity Site, New Mexico, USA	226
Nagasaki, Japan	228
Hiroshima, Japan	230
Mekong River, Vietnam	232
Mesopotamian Marshes, Iraq	234
Brandenburg Gate, Berlin, Germany	236
Mostar Bridge, Bosnia	238
Beirut, Lebanon	240
Qalqilya, Palestine	242
Twin Towers, New York, USA	244

Leisure and Culture

Mount Rushmore, USA	248
Luxor Temple, Egypt	250
Bamiyan Valley, Afghanistan	252
Mayan Temple, Central America	254
Cleopatra's Needle, Egypt/England	256
London Bridge, England/USA	258
Great Wall, China	260
Qinghai–Tibet Railway, Tibet	262
Woodstock, USA	264
Zermatt, Switzerland	266
Zugspitze, Bavaria, Germany	268
Cornwall, England	270
Shenzhen, China	272
Dal Lake, India	274
Paris, France	276
Jumeirah Beach, Dubai, United Arab Emirates	278
Benidorm, Spain	280

Rising tide The Pacific Islands nation of Tuvalu will probably disappear beneath the waves sometime during this century, thanks to global warming. Here, in February 2002, the waves from its lagoon eat away at the shore right in front of the country's environment ministry.

Foreword

According to an exhaustive 2007 survey by the Intergovernmental Panel on Climate Change, we can be 90 percent certain that humans are responsible for the climate change we've seen in the past 50 years. The sheer extent of the change the panel reports is truly alarming. Moreover, its warning is echoed by countless scientific authorities, including the World Meteorological Organization and all the National Science Academies of the G8 nations, and it adds yet another chapter to an already-impressive body of scientific evidence of our impact on the planet.

But, as ever, words often fall desperately short of capturing the true magnitude of the transformations we're currently witnessing. The public has been bombarded with data, predictions, and statistics relating to the environment for years, and yet somehow environmental science still can't avoid appearing abstract and distant for so many. That's why this extraordinary book is so important.

Whoever said that a picture can replace a thousand words clearly had *Earth Then and Now* in mind. Its stark visual comparisons speak volumes, and, as a reporter on environmental issues for the past 16 years, Fred Pearce is uniquely placed to tell the stories behind the images. There are few people better qualified than he when it comes to explaining the major issues facing our planet.

Guiding us through the changes the Earth has undergone over the past hundred years, he expertly highlights the key issues affecting the world today – environmental change, urbanization, land transformation, the forces of nature, war and conflict, and leisure and culture. The juxtaposition of "before" and "after" photographs forces us to look at often-devastating changes – and, in parts, exposes us to the darker side of mass affluence. Images of landslides, volcanoes, and earthquakes, meanwhile, provide us with a humbling reminder of nature's immense power. Perhaps the most shocking example of this is Katrina: a single hurricane that caused unprecedented devastation.

Each pair of photographs tells a powerful story. Some – such as the large-scale melting of the polar ice caps – are global, while others – like the Stieregg restaurant in Switzerland dangling precariously on the edge of a precipice – are more local. The images stretch across time, ranging from the devastation of the Somme battlefield in World War I to the phenomenal urban expansion of Hong Kong in the 1990s.

Earth Then and Now is a deeply unsettling yet hugely compelling collection. It reinforces what we instinctively know, and what a growing scientific consensus is telling us: that climate change is the biggest threat we face. In the light of this evidence it's our job, as individuals, as consumers with choices, as investors, and as voters, to apply whatever pressure we can for change. This book shows us how the Earth was, and it reveals what it has become. But what it cannot do is show us the future. That is up to us.

Zac Goldsmith
THE ECOLOGIST

September 2007

Introduction

Two years ago I stood on the promenade at Muynak, an old seaside resort on the shores of the Aral Sea in central Asia. Behind me was a fish-processing factory that had once sent canned fish across the Soviet Empire, from Warsaw to Vladivostok. No longer. Looking out to what should have been sea, I saw fishing boats abandoned on a beach that went on forever. The sea had disappeared more than 30 years ago, and its bed had turned into a new and entirely unexplored desert. Over the horizon, I was told, a remnant of what had once been the world's fourth-largest inland sea remained. But the shore was more than 100km (60 miles) away now and nobody in the town had ever gone to see it. I spotted a fox trotting where fish had once swum. In the far distance, a dust storm was brewing.

The Aral Sea died because Soviet engineers removed all the water from the two great rivers that once kept it full. They took the water to irrigate vast expanses of cotton farms – to grow the uniforms for the Red Army. The Russians have been gone for more than a decade now, of course, but the abstractions continue, and the fields are today growing cotton for shirts and jeans and underwear on sale in almost every high street in the world. We are to blame now, as the sea continues to evaporate in the heat of the desert sun. It will probably be entirely gone within a decade.

UN scientists call the emptying of the Aral Sea the greatest environmental disaster of the 20th century. But I only really understood the scale of what had happened when I returned home from Muynak and looked at the pair of satellite images that appear in this book. They show a whole sea reduced to a toxic sump by human action. It is an unprecedented man-made change to the shape of the world. This book is full of such pairs of images, showing how our world has been changed. Most show bad things we have done, but by no means all. There are good stories here, too. This is the latest chapter in man's use and abuse of planet Earth. But all the images beg the question: what will happen next?

Humans have been making their mark on the landscape for a long time. Our hominid ancestors discovered how to make fire more than a million years ago, and they hunted animals across the plains of their first home, Africa. By the time of the last ice age our own species, *Homo sapiens*, had learned to combine these two skills to good effect, setting huge fires across the grasslands to drive animals towards their spears. It was the start of what has probably been mankind's biggest and most destructive impact on the land surface: deforestation. But we soon ceased to be hunters and gatherers alone. For at least 10,000 years, since the end of the last ice age, we have been farming. To make way

Once, this was one of the great fishing grounds of the Soviet Union. But more than 30 years ago the waters disappeared over the horizon, the fish were gone, and the trawlers left high and dry. It was as if a giant plug had been pulled in the bottom of the Aral Sea. In fact, the taps had been turned off. For the purposes of commercial irrigation, engineers were taking almost the entire flow of the two rivers that topped up the inland sea – and evaporation in the hot sun has done the rest.

A drop in the desert
Wadi Rum is the largest wadi in Jordan. Lawrence of Arabia fought here, and the film of the same name was shot here. Engineers now pump water from beneath the desert, and pivot sprays irrigate circular patches of sand to grow grain. But the water is running out. Soon the desert will reclaim its own.

for the crops needed to feed a fast-growing human population, we chopped and burned down ever-greater areas of forests, ploughed the great roaming grounds of wild animals, and drained huge areas of marshes. Nature was on the retreat.

But, thanks in part to the ephemeral nature of many human settlements, nature has demonstrated a remarkable ability to recover. Many apparently natural forests contain evidence that ancient human civilizations, of which we know virtually nothing, cleared them thousands of years ago. Afterwards, nature returned. Modern explorers have rediscovered the great Mayan ruins, for instance, buried in thick jungle in Central America. And those ruins are far from unique. Six hundred years ago there were cities with substantial populations in the Amazon jungle. Go back 1,500 years and the forests of central Africa were being turned into charcoal for metal smelting.

So, the good news is that nature can recover from the impact of human activity. But the bad news is that she has never experienced anything like the intensity of our current interventions. Today there are almost seven billion humans on the planet, a thousand times more than 10,000 years ago. And we now have advanced technology at our disposal. Once, we damaged small areas and then moved on; now very little of our planet is untouched by human occupation, and often the damage looks terminal.

Photographs can not document those early footprints of humanity, but they do cover the past 150 or so years, during which our population and our impact has soared. In these pages you will see some of the consequences. In that time we have chopped down half the world's forests, doubled the area of land under the plough, all but eliminated large mammals, drained the majority of the world's marshes, and reduced most of the

oceans' fish stocks and whale species by more than 90 percent. We have paved huge areas of the planet and broken much of the rest of the natural environment into tiny fragments by our extensive road-building. We have re-plumbed the rivers, plugging most of them with dams so that many no longer deliver water to the sea, and we have diverted their water instead through thousands of canals to irrigate fields. The demise of the Aral is only one – albeit the worst – outcome of that hydrological plunder.

We have dug deep into the bowels of the Earth. Look at the pictures of what we have done to Bingham Canyon in the USA and to Chile's Atacama desert – all in the pursuit of copper. And consider the way in which we have mined the South Pacific paradise of Nauru until there is, more or less literally, nothing left. Meanwhile, by releasing chemicals into the air we have burned a hole in the ozone layer and fundamentally altered nature's methods of recycling key elements such as sulphur, nitrogen, and – perhaps most dangerously for the future habitability of our planet – carbon.

Even seen from space, our handiwork is glaringly obvious. It is visible in huge reservoirs and megacities, in land drained from the sea, in whole coastlines that shift under our hidden hand, in dust storms that circle the globe, and in disintegrating ice sheets.

And yet there are things we can be hopeful about. Who can fail to be moved by the architectural magnificence of the Millau viaduct in southern France, or by some of the dazzling modern cityscapes, or even, for its sheer *joie de vivre*, by the Paris beach? The construction of the Panama Canal killed thousands, but it remains an extraordinary monument to entrepreneurial and engineering endeavour. We should marvel, too, at the civil determination behind the rebuilding of San Francisco after the 1906 earthquake, the war-damaged Frauenkirche in Dresden, and the new Mostar bridge in Bosnia.

Our creativity and inventiveness may have got us into our current environmental mess, but we must hope that they can get us out of it too. And there is evidence that they can. Look at some of the environmental rehabilitation projects undertaken after the closure of old mines. And think how the bleak, smog-filled industrial heartland of Bilbao has been transformed by an imaginative cultural vision. Scars can be healed.

Despite this, there is no denying that these chapters also contain evidence of tragedy, stupidity, venality, and short-term thinking in abundance. And, sadly, it is these misdemeanours that, for the time being at least, hold sway over the various examples of rehabilitation. Why, for example, care about coastal mangroves when they can be turned into toilet paper and the land annexed for prawn farms? Why worry about drying up the Aral Sea when there is cotton to be grown? Such human willfulness has, of course, been immensely compounded by advancing technology. We have shovels that can lift 100 tonnes of ore in one scoop and dig holes up to a kilometre (3,300ft) deep. We plug rivers

By the time you read these words, this town will be no more. Wanzhou, on the banks of the River Yangtze, is being drowned by the reservoir now filling behind the mighty Three Gorges dam. The sign on the building marks 175m (575ft), the reservoir's planned water level. To prevent the dam's hydroelectric turbines getting clogged, workmen are removing every trace of the town before the floodwaters rise.

with concrete barriers hundreds of metres high, and drain marshes and defoliate rainforests because we can. Sometimes the scale of our endeavours is overwhelming. In Tokyo, for example, we have paved an area of more than 5,000sq km (2,000sq miles).

But we don't just inflict damage on nature. We do it to each other, out of fear and hate. The wars of the 20th century, I sincerely hope, will be looked back on by future generations as outbreaks of collective madness, never to be repeated. Thanks to the industrialization of warfare tens of millions of people have died, often for causes that were illusory or have long since ceased to matter. For me, trying to sum up in a couple of hundred words what exactly happened, and why, at Passchendaele – an event nearly a century ago – was almost unbearable. The devastation of Hiroshima and Nagasaki and the siege of Leningrad also chill the soul. But perhaps the worst is over. Horrific as modern terrorist outrages such as the destruction of the Twin Towers are, they are orders of magnitude smaller in scale than the state-sponsored carnage of the two world wars. Nuclear weapons have not been used in anger since the summer of 1945. But, of course, while they continue to exist we remain under their shadow.

Nuclear weapons famously claimed to unleash the power of the sun. And, for all our inventiveness and technological sophistication, the forces of nature remain supremely powerful influences on the habitability of our planet. At times we have simply to stare in wonder: at the power of floods to wash away our world; at the force of earthquakes as the Earth's crust shifts; at tsunamis and avalanches that wipe out tens of thousands of people in their path; and at volcanoes that throw in our faces the molten contents of the Earth's core and shroud the planet in dust. One such volcano, 73,000 years ago, cooled the planet so much that our species came close to being wiped out.

And yet the truth is that we humans are beginning to change these great geological and planetary processes. Witness the extraordinary images of how the ozone hole opens up over Antarctica each southern spring. This is an entirely man-made phenomenon. The chemicals that make it happen did not exist in nature before we invented them in the early 20th century. Discovering the ozone hole in the mid-1980s was the moment when scientists realized that we truly could accidentally destroy our world.

And witness, too, the many images of melting glaciers and collapsing ice sheets. The Earth's stores of coal and oil and gas are the remains of carbon that nature captured from the air in much hotter times and gradually buried over tens of millions of years. That capture helped cool the Earth and keep it fit for the rise of humans. But now, by digging up and burning those fossil fuels, we are releasing that carbon back into the atmosphere. Every year, we unleash carbon that it took nature a million years to bury. Few people, beyond the perverse, can be truly surprised that our planet is warming fast as a result.

Known to the world's air travellers simply as JFK, New York's John F Kennedy International Airport is one of the lynchpins of the global flying business. It has no fewer than nine terminals and handles more than 40 million passengers a year. Most visitors to the USA arrive on its runways. It dispenses more than four billion litres of fuel a year to aircraft preparing to take off, making it one of the world's global-warming hubs.

Allied bombing destroyed most of the centre of Cologne, Germany's fourth-largest city, in 1942. The twin spires of its famous Dom were among the few structures to survive. Behind, the Hohenzollern Bridge across the River Rhine lay badly damaged.

I am not among those who believe that we are doomed either by the pollution we have created or by our ransacking of the planet's resources. But nor am I indifferent to what we have done, as a significant minority of people still seems to be. The stakes are high. We do have the power to destroy our world. Nevertheless, I hope that you will see in these images some signs of our virtue as well as of our sin; of reasons for hope as well as for despair.

For me, one of the strangest pairs of images is that of Pripyat – a Soviet town built to house the workers at a power station in the Ukraine. The first of the two pictures looks like any other bleak landscape of tower blocks built by unthinking bureaucrats. The second image is, at first sight, similar, until you notice all the overgrown trees and weeds in the squares and streets. And the absence of people. It is like a post-apocalypse landscape.

And that, for the people of Pripyat, is what it is. For they worked at Chernobyl, the nuclear plant that caught fire in 1986. It released so much radioactivity that the town was evacuated and will not be safe enough to reoccupy for centuries hence. This is a human disaster, of course. But it has been a boon for wildlife. Down there among the trees are wild boar and deer, elks and wolves. They may be a bit radioactive, but they have revelled in the departure of humans and reoccupied the city. Nature has returned, just as it did after the fall of the ancient rainforest empires.

Is there a final lesson here? I think so. Nature is not as fragile as we think. She is resilient. With time, she may recover from the worst we can throw at her. It is we, ultimately, who are the fragile ones. Look at these pictures and fear not so much for nature: fear for us.

Environmental Change

Where does the wrath of nature end and the influence of mankind begin? As we watch deserts grow and ice caps melt, it is not always obvious. Nature is always on the move; never a passive spectator. But even so, this chapter reveals growing evidence that humans are implicated not just in obvious things, such as chopping down rainforests and draining wetlands, but also in apparently natural changes to the landscape.

Several of the examples involve rivers. Rivers flood naturally, of course. Nothing new in that. But many floods are as much to do with human activity as nature. In nature, rivers have floodplains over which they freely roam, but we have taken over much of this land, barricading it off from the rivers. No wonder, then, that great rivers such as the Elbe in Germany seek out the weaknesses in our flood defences and overwhelm them, causing floods far more destructive than anything that came before.

All rivers can suffer from low flows, too. But river abstractions, such as on the River Yangtze in China, can artificially empty rivers. And engineering on rivers such as the Rhine, intended to make it more navigable, has contrived to increase the risk of both low and high flows. Likewise, all rivers erode silt from mountains and carry it to the sea. But when humans are involved in deforestation of the hillsides, the silt flow increases. We see the consequences in northern China, where the Yellow River delta is being remade.

All these changes are dramatic, but local. Mankind's newest and most pervasive impact is global climate change. Nobody now doubts that the world is warming. And few outside a hardened band of climate sceptics deny that the billions of tonnes of heat-trapping greenhouse gases that humans have put into the atmosphere are largely to blame.

In the past 30 years, for instance, climate change has doubled the area of the planet suffering from droughts. Combine that with bad environmental practice on the ground and you have a major crisis. In West Africa, reduced rainfall and water abstraction for irrigation projects has almost dried up Lake Chad. In northern China, drought and bad farming is eroding soils and creating massive dust storms that infest the capital and blow all the way to Canada. In the Russian far east, drought has turned forests to a tinderbox.

The most unambiguous evidence of warming is the near-universal retreat of the world's glaciers. Among the examples we show here, the most famous is right on the equator, where Kilimanjaro, Africa's tallest mountain, is rapidly shedding ice. Within a decade it could be ice-free. The great polar ice caps, which have the potential to raise sea levels by tens of metres if they melt, are also under pressure from global warming. This chapter includes examples from Antarctica, and evidence that Arctic sea ice, too, is in full retreat and could be gone altogether in summer within a few decades. We are starting to see the consequences of all this in rising sea levels all around the world. From St Mark's Square in Venice to the paradise of the Maldives in the Indian Ocean, *terra firma* itself is under threat.

The River Yangtze in southern China is better known for its disastrous floods, but in summer 2006 this major tributary, the Jialing, dried up. Here is its dessicated bed as it enters the fast-growing megacity of Chongqing, just before it joins the Yangtze. Children pick their way over the cracks. The river died in part because rains failed in the river's headwaters. Much of southern China was suffering the worst drought in a century. But in recent years, too, farmers have taken ever more water out of the river to water their rice crops. That turned low flows into no flows.

TRIFT GLACIER
Switzerland

2002 The Trift Glacier is in the central Swiss Alps. A favourite with mountaineers, it carries ice formed on the 4,000m- (13,000ft-) high Wiessmies Mountain, and ends in a long tongue of ice that crosses a lake that was gouged out by the glacier during the last ice age. The tongue has been a permanent feature of the glacier since it was first scaled in 1855, and probably for much longer.

Retreating ice Only one year after the first picture was taken the ice tongue has been replaced by open water. The ice had retreated by more than 500m (1,640ft). A permanent path across it, used by generations of mountaineers, has disappeared. The melting was caused by exceptionally warm weather during the summer of 2003, which triggered a massive thaw right across the Swiss Alps. Glacier tongues such as this one proved particularly vulnerable in the face of a combination of warmer air and warmer water. Glaciologists say the Trift tongue is unlikely ever to reform. After this picture was taken tour operators replaced the path with a suspension bridge across the lake. Meanwhile, filled with meltwater, the rising lake is close to breaking its banks, and engineers have been siphoning water in order to lower water levels and prevent a catastrophic tidal wave shooting down the valley.

CORDILLERA BLANCA
Peru

1980 The Cordillera Blanca is a mountain range in Peru that forms part of the larger chain of the Andes, which runs the length of South America. In the cold, rarefied air here, there is more tropical ice than anywhere else on the planet. And in Jacabamba valley, on the remote eastern plank of the Cordillera and surrounded by snowy peaks up to 6,000m (19,000ft) high, this huge glacier edges down the mountainside into a lake at the bottom, from where it was photographed by British geologist Bryan Lynas. But the lake, and the broken ice in it, is a sign that melting is already under way.

Disappearing glacier Just over two decades later the Jacabamba glacier is a shadow of its former self. A few patches of ice persist in the darkest, most shaded part of the valley bottom, but otherwise the mountainside is now bare rock. The right-hand photograph was taken 23 years after the first, from exactly the same spot as the earlier one, by Lynas's son, the author Mark Lynas. Across Peru a quarter of the ice has disappeared in the past 30 years. Global warming is to blame.

GRINNELL GLACIER USA

1940 George Bird Grinnell was one of the first Europeans to explore the wide-open spaces of Montana in the foothills of the Rocky Mountains. Unlike most other visitors, who were searching for gold and minerals, he was struck by the area's scenery, and especially its glaciers. There were no other glaciers in the surrounding mountains, and this corner of the state stood out like an ice island. He campaigned for the creation of the Glacier National Park here, and, after the Park's formation in 1910, one of the glaciers, on the slopes of Mount Gould, was named after him. Back then, Grinnell Glacier covered more than 2sq km (0.7sq miles). But soon a lake of meltwater started to form at its base.

Montana meltdown
Today Grinnell is less than half the size it was at the start of the 20th century. It is now broken into six patches, each disappearing fast. The lake has grown dramatically, and is often littered with small icebergs. The Glacier National Park still has about 50 glaciers. During the 1960s and 1970s – a period of heavy snowfalls – several of them grew, but since serious warming began in the area around 1980 all have been in retreat. In the century since the Park was established, three-quarters of the area once covered by glaciers has become lakes or bare rock. At the current rate they will all be gone by 2030. Then the park will need a new name.

RHÔNE GLACIER Switzerland

1932 The Rhône Glacier once filled the Grimsel Pass in the Swiss Alps, on the southern slopes of the Dammastock Peak. The glacier is a vital resource, providing meltwater that fills Lake Geneva 150km (95 miles) downstream, and forming the starting point for one of Europe's most vital navigation arteries and sources of hydroelectricity, the River Rhône. The glacier has been slowly retreating since the last ice age ended 10,000 years ago. Here, in the 1930s, it is less extensive than in the mid-19th century, when it reached right to the town of Gletsch in the valley bottom.

Unstoppable thaw Since the 1930s the glacier has been disappearing ever more rapidly, largely as a result of the significant warming in the Alps. Particularly in the past decade, its retreat up the mountainside has accelerated sharply. Local towns depend on the glacier to attract visitors, and each summer, now, the people of Gletsch rush out to cover the glacier with huge insulating blankets to try to protect its tongue from further melting. But to little avail. The retreat goes on. In the short term the thaw has increased river flow, but in the long run, as the glacier comes close to disappearing altogether, the regular supply of water from it will disappear. Then Lake Geneva and the River Rhône will be entirely dependent on rainfall.

UPSALA GLACIER Patagonia

1928 Eight decades ago the glaciers of Patagonia in southern Argentina were mostly advancing. Among them was South America's largest, the Upsala Glacier. It was named after the Swedish city whose university carried out the first surveys of the glacier, although this picture was taken in 1928 by one of the earliest Argentinian explorers of the Patagonian glaciers, Mar'a Albert of Agostini, "the Father of Patagonia". The glacier's river of ice flowed down from the mountains into the Argentino Lake, which it blocked and filled with its icebergs.

Open water The glacier has retreated. It no longer blocks the Argentino Lake, and what was once ice has turned to water. Upsala Glacier is now almost 4km (2.5 miles) shorter than in 1928. The retreat has been unrelenting since 1978; the glacier lost almost 0.8km (half a mile) in 1994 alone. Global warming has almost certainly played a part in this, but the change may have been accelerated because less ice is flowing down the glacier today. Nonetheless, all told, the Patagonia glaciers and ice fields have lost 2,000cu m (70,600cu ft) of ice every year for the past seven years – a faster rate of loss than anywhere else on Earth.

STIEREGG RESTAURANT Switzerland

2004 The Stieregg restaurant had been in business since 1952, more than 1,600m (5,200ft) up in the Swiss Alps. It was a regular destination for people heading towards the Grindelwald glacier on the slopes of the Eiger. But the glacier has been in retreat for some years. And, more critically still, the permafrost beneath it is melting. Many mountainsides like this are held together by the permanently frozen soil. But as temperatures rise the soil melts and what was once as solid as rock turns into an unstable scree slope. It doesn't take much to turn that slope into an avalanche as lethal as any made of snow.

Mountain erosion Rising temperatures have sealed the fate of the Stieregg restaurant. First the thawing mountainside above it began to unleash boulders down the slope. Millions of tonnes of rock have broken off the side of the Eiger in recent years. Then, during the height of the summer melting season in 2005, the swollen river in the valley beneath the restaurant began to erode its banks. At the end of May, some 300,000cu m (10,600,000cu ft) of thawed soil crashed into the valley and the restaurant was left hanging over the edge of a rapidly eroding precipice. The local authorities eventually set fire to the building to prevent it from adding to the debris below – another victim of global warming.

MOUNT KILIMANJARO Tanzania

1974 There has been ice on top of Mount Kilimanjaro for at least 12,000 years. Although the mountain is only 350km (220 miles) from the equator, the air at its 6,000m (19,000ft) summit has been cold enough to turn the rain to snow and keep it frozen all year. The snow and Kilimanjaro's status as the highest peak in the world accessible to ordinary tourists have made the mountain Tanzania's top tourism moneyspinner. But Kilimanjaro, which means "shining mountain", is losing its sparkle. The ice fields in its volcanic crater and the glaciers down its flanks have been disappearing since they were first measured in 1912. When this picture was taken in 1974, the ice was already in retreat.

Melting ice cap Today the ice is disappearing faster than ever. Early ice loss was probably down to declining snowfall caused by deforestation on the slopes. But today, warmer temperatures are melting the slopes fast. More than 80 percent of the ice is now gone, and on current trends the rest will melt away by 2015. Around the world, tropical glaciers are disappearing – from nearby Mount Kenya to the Peruvian Andes and the peaks of New Guinea. They say you can climb Kilimanjaro in trainers now. But, paradoxically, the mountain is becoming more dangerous, because the disappearance of ice and melting of permafrost are loosening rocks and causing more and more landslides.

FILCHNER ICE SHELF Antarctica

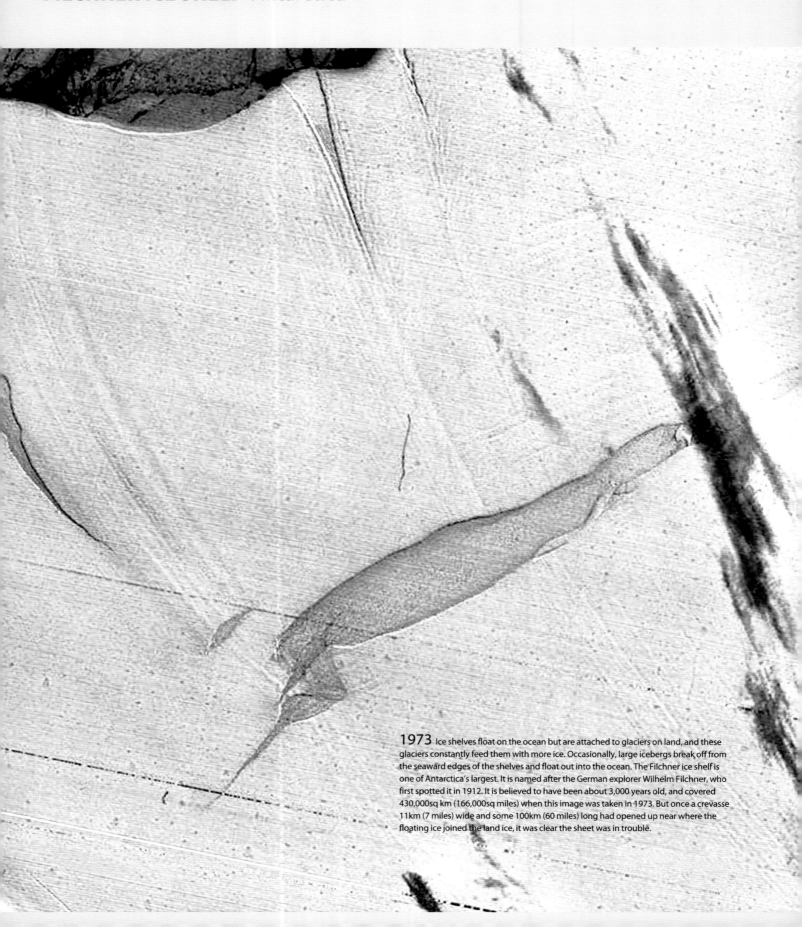

1973 Ice shelves float on the ocean but are attached to glaciers on land, and these glaciers constantly feed them with more ice. Occasionally, large icebergs break off from the seaward edges of the shelves and float out into the ocean. The Filchner ice shelf is one of Antarctica's largest. It is named after the German explorer Wilhelm Filchner, who first spotted it in 1912. It is believed to have been about 3,000 years old, and covered 430,000sq km (166,000sq miles) when this image was taken in 1973. But once a crevasse 11km (7 miles) wide and some 100km (60 miles) long had opened up near where the floating ice joined the land ice, it was clear the sheet was in trouble.

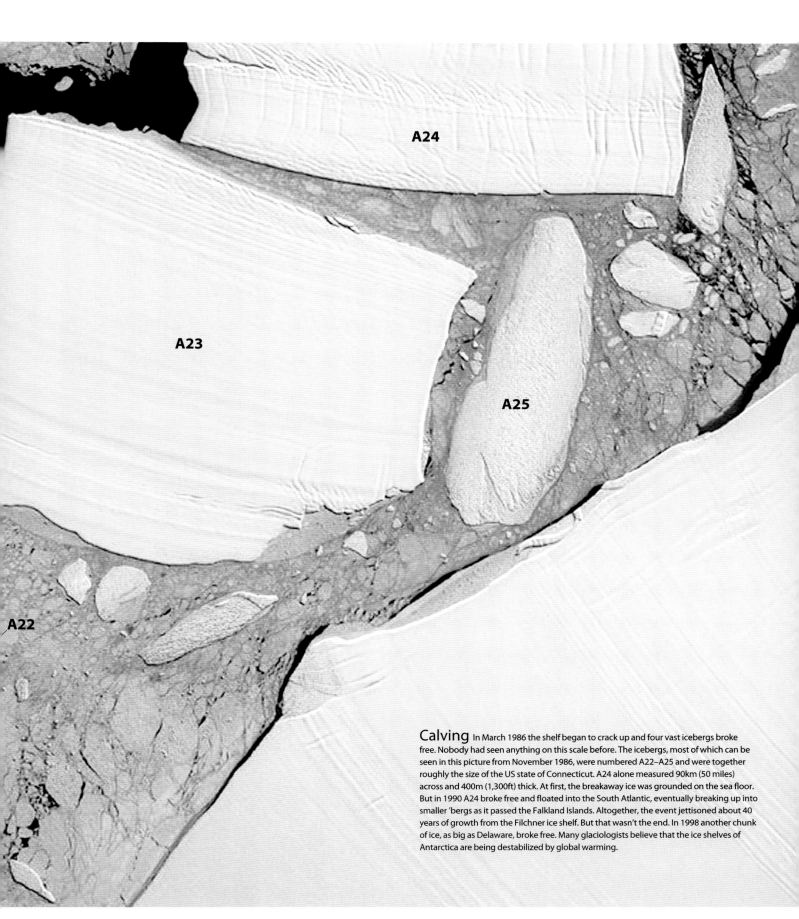

A24

A23

A25

A22

Calving In March 1986 the shelf began to crack up and four vast icebergs broke free. Nobody had seen anything on this scale before. The icebergs, most of which can be seen in this picture from November 1986, were numbered A22–A25 and were together roughly the size of the US state of Connecticut. A24 alone measured 90km (50 miles) across and 400m (1,300ft) thick. At first, the breakaway ice was grounded on the sea floor. But in 1990 A24 broke free and floated into the South Atlantic, eventually breaking up into smaller 'bergs as it passed the Falkland Islands. Altogether, the event jettisoned about 40 years of growth from the Filchner ice shelf. But that wasn't the end. In 1998 another chunk of ice, as big as Delaware, broke free. Many glaciologists believe that the ice shelves of Antarctica are being destabilized by global warming.

LARSEN B ICE SHELF Antarctica

January 2002
The Larsen B ice shelf was an area of floating ice roughly the size of Luxembourg and connected to a peninsula of Antarctica that juts out into the South Atlantic. Glaciers in the mountains visible on the left of this image fed ice onto the shelf. Small icebergs would periodically break off from its outer edge, but the shelf itself had existed for at least 10,000 years, and looked secure enough when this picture was taken on 31 January 2002.

Shattered
Appearances can be deceptive. The Antarctic peninsula, and the waters around it, had been getting warmer. The shelf had been melting from beneath, and pools of liquid water had formed on its surface. In mid-February 2002 – the warmest time of year in the southern hemisphere – the ice abruptly shattered. By 23 February a series of huge icebergs could be seen breaking away, and the shelf coastline had retreated.

New shoreline
By 7 March most of the floating ice shelf had disintegrated. Numerous huge icebergs had begun heading out to sea, and the light-blue area was a large new bay full of the shattered remains of the ice shelf, gradually melting. In all, 2,600sq km (1,000sq miles) of ice shelf disappeared.

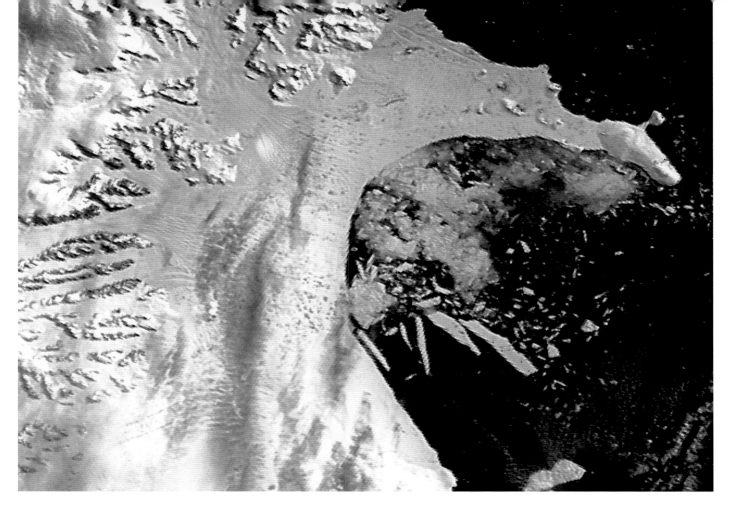

ARCTIC OCEAN North Pole

1979 One of the most pronounced effects of global warming over the past three decades has been the reduction in permanent Arctic sea ice. This satellite image from September 1979 shows the sea ice at its minimum during the first year in which NASA collected satellite images. This ice has lasted through the summer and will form the basis for the accumulation of new ice during the winter. The ice extends across virtually all of the Arctic Ocean north of 75 degrees, except for some areas north of Europe and Asia that are kept warm by the Gulf Stream. And it pushes south of 75 degrees through the islands of the Canadian Arctic and along the east shore of Greenland.

Arctic meltdown By 2005, when this photograph was taken, the picture looks very different. There is a marked change in the distribution of the permanent ice. It has, for instance, retreated from many of the passages and gulfs of the Canadian Arctic and from large areas of ocean north of eastern Siberia, while encroaching closer to the Norwegian islands of Svalbard. Overall, there has been a reduction in the total area of sea ice of more than a fifth, with 2005 showing the smallest ever recorded ice cover.

LAKE DONGTING China

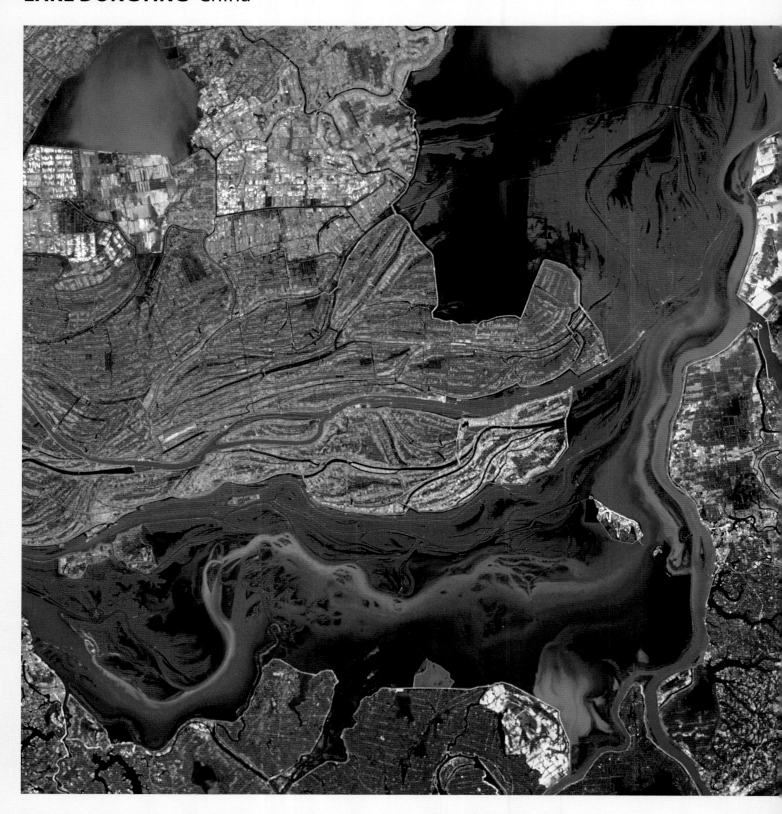

May 1998 Lake Dongting is a permanent feature on the floodplain of the river Yangtze in central China. The lake (seen in blue in this infra-red image) covers 2,700sq km (1,000sq miles), an area larger than that covered by Greater London, England. But it is less than half the size that it was a century ago. Since then, farmers have drained huge areas of former lake bed for rice paddy and much of the rest of the lake has filled with silt brought down by the Yangtze from deforested and eroding hillsides upstream in remote areas of western China. Most of the silt is covered in vegetation and shows up in the picture as red.

Sudden flood Each summer the monsoon rains swell the Yangtze, and farmers around Lake Dongting face the risk of inundation. But in 1998 the flooding was catastrophic. Between the first picture, taken in May, and this second in August, the lake has grown to several times its normal size, engulfing rice paddy and other low-lying areas. The flood killed 3,000 people and destroyed five million homes. Its severity was partly due to exceptional rains, but partly also to the denuded hill upstream that can hold less rainwater today than in the past. After the 1998 floods, in an attempt to prevent a repetition, China banned all logging of natural upland forests.

VENICE Italy

2001 St Mark's Square attracts the biggest crowds in one of the world's most popular tourist cities, Venice. Millions flock here to see the ancient basilica and sit at café tables in the piazza. Venice is unique. The former capital of a hugely successful merchant empire, it is constructed on wooden piles in a lagoon at the head of the Adriatic sea. There are no roads or cars here, only boats travelling the ubiquitous canals that are its thoroughfares. But the water all around the city is also its biggest threat. And St Mark's Square, located right by the Grand Canal, is the city's lowest point.

High water Venice has always been at risk of flooding, but man-made changes to the lagoon, and rising global sea levels, have greatly increased the danger. Once, the lagoon was rarely more than a metre deep, and its mud banks absorbed the force of the waves, keeping out high tides. Today waves surge into the lagoon along shipping channels dredged for cruise ships and other vessels. Many of the mud banks have been eroded. In the past three decades the average high tide in the lagoon has risen by almost half a metre, flooding basements and washing across St Mark's Square around 50 times a year.

YELLOW RIVER China

1979 The Yellow River is the muddiest waterway on earth – hence its name. Every tonne of water contains 40kg (90lb) of silt. Most of that silt is washed into the ocean, but some of it drops to the river bed before it reaches the sea, creating the river's large and growing delta. Most of the land in this image is delta. In recent decades, a combination of accelerated erosion upstream and slower river flows caused by dams has ensured that ever more of the silt ends up in the delta. The large brown bulge of silt on the east side of the delta shows the most recent area of growth, along the river's current main route to the sea.

Yellow River

Extending delta Two decades on, the cumulative impact of all that silt on the delta is clearly visible. The bulge has turned into a giant beak around the river's main channel. And a secondary beak is forming around a second channel. Between 1979 and 2000, several hundred square kilometres of new delta have formed. The silt will make good farmland, but the risk is always that it will one day block the channel and force the river to make a sudden and catastrophic change of course. The last time that happened, back in 1938, 900,000 people died in the ensuing floods and famines.

Yellow River

HITZACKER Germany

1999 The medieval German town Hitzacker sits on the River Elbe where it joins the River Jeetzel in Lower Saxony. It is a beautiful setting. This small settlement, with its many half-timbered buildings, is surrounded by extensive flood meadows that are a paradise for birdlife. Many of its 5,000 inhabitants can have boats moored close to their houses. Around town, however, are signs of a more troublesome side of this fluvial idyll. Marks everywhere reveal that high water often enters the town, flooding its cellars and alleyways.

Frequent flooding

Inundation of this sort is happening more and more at Hitzacker. In August 2002 CNN came to the town to film exceptional floods rushing down the Elbe, but this picture was taken four years later in April 2006, when Hitzacker was again swamped after heavy rains combined with snowmelt in the Krkonose Mountains of the Czech Republic to swell the river. The chancellor, Angela Merkel, toured the town as the Elbe hit the highest marks ever recorded. Environmentalists blamed dykes and developments upstream that cut the river off from its floodplain and pushed river levels ever higher. Vulnerable towns such as Hitzacker paid the price.

LAKE CHAD Africa

1972 Lake Chad is a giant evaporation pond that drains the southern fringes of the Sahara desert. The lake is extremely shallow, averaging only 2m (6ft) or so deep. So its size changes dramatically as rainfall fluctuates in its catchment area. It was for a time the world's sixth-largest inland sea and straddled parts of four countries: Chad, Cameroon, Nigeria, and Niger.

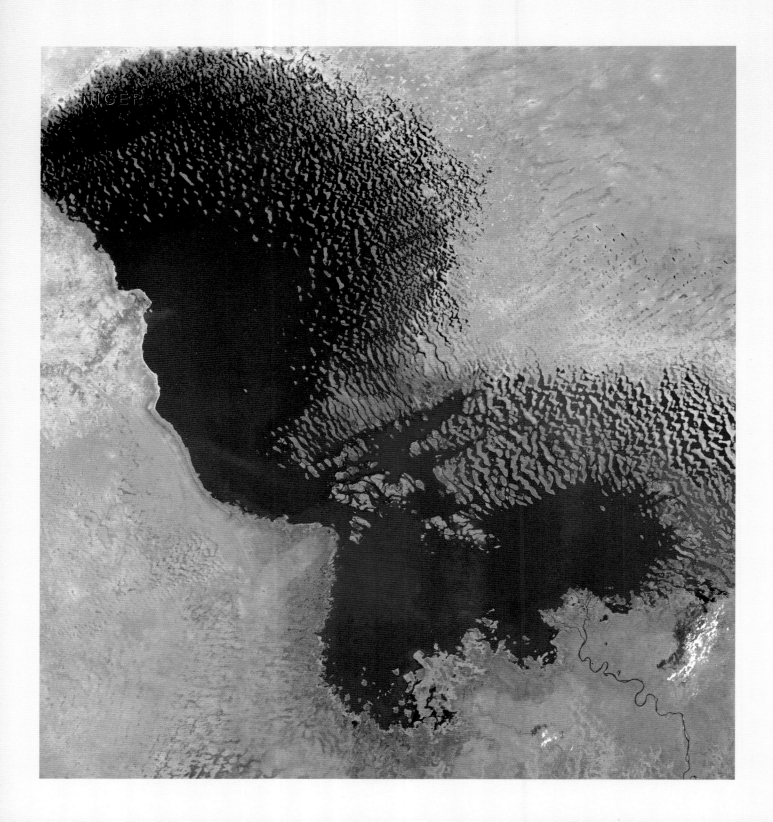

Drying out But in recent decades the rains in this region have been fitful. River flows into the lake have declined. Meanwhile, big irrigation projects have taken water from the rivers, reducing their flow further. Lake Chad has been drying out and its shoreline has retreated from Nigeria and Niger altogether. Traditional farmers around the lake have long since learned to cope with this constant change. They simply move with the lake edge. But large, engineered irrigation systems have suffered. The British-designed South Chad Irrigation Project in Nigeria, which was built in the 1960s to take water from the lake, is now high and dry, with its lowest intake tens of kilometres from the nearest shore.

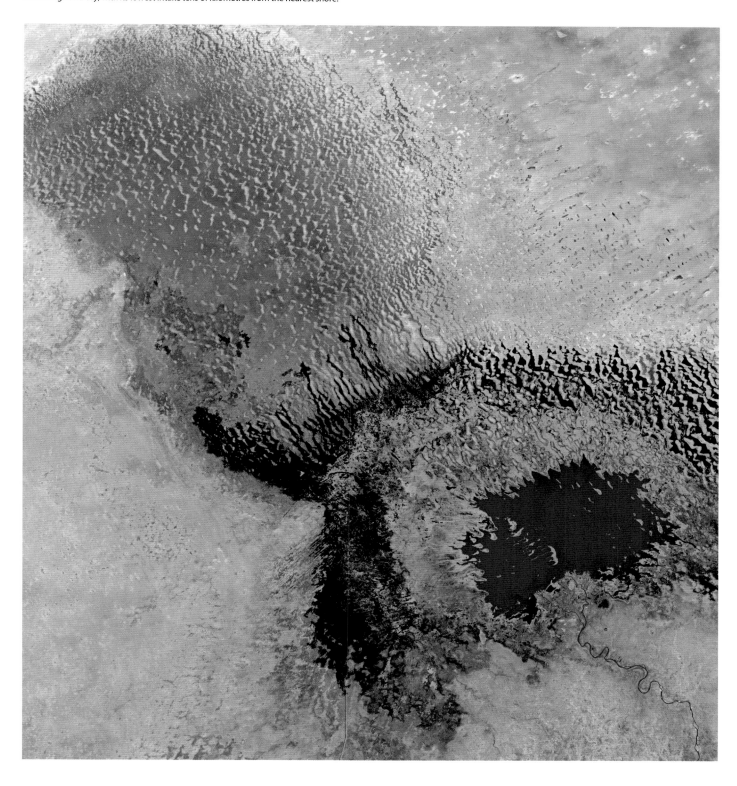

RIVER RHINE Germany

1990s Dusseldorf is one of the great German industrial cities. Once a small market town on the banks of the River Rhine, it has grown into a large and prosperous metropolis since the river was re-engineered in the 19th century as a single waterway navigable all the way from the North Sea to the Swiss border. In good times the modern engineering works well. Giant barges and even ocean-going ships make their way swiftly up and downstream on their way to the Ruhr industrial region, for which Dusseldorf is the financial centre. In this image the river is full, providing a fine backdrop for the old town, with the city's television tower overseeing everything in the background.

River drought The new river may be generally more navigable, but it is also more capricious. Much of its floodplain upstream has been drained and barricaded from the river. There are no wetlands to act as sponges, absorbing flood flows and keeping the river flowing during droughts. The result is more periods of high downstream flows in cities such as Dusseldorf, where peak river flows are today a third higher than a century ago. But there are also more frequent episodes of low river flows, and the second picture shows one such, during the European drought and heatwave of 2003, when the river virtually dried up for several weeks.

ARAL SEA Kazakhstan/Uzbekistan

1973 The demise of the Aral Sea in Central Asia is one of the greatest environmental disasters of the 20th century. Back in the 1960s, it was the fourth-largest inland sea in the world and had one of the Soviet Union's biggest fishing fleets. But at Muynak, the main Uzbek port, the sea disappeared over the horizon in the early 1970s. By the late 1980s the fish had disappeared and areas of salty desert were appearing where once there had been breaking waves. The great rivers that once filled the sea were no longer delivering water because Soviet engineers were abstracting most of their flow to irrigate endless fields of cotton. With no new water, the sea was slowly dying in the desert sun.

Kazakhstan

Uzbekistan

Shrinking sea
Today Moscow's cotton commissars have long since departed. But the independent nations of Uzbekistan, Kazakhstan, and Turkmenistan are still emptying the rivers to grow cotton for Western markets. These three countries have the highest rates of per-capita water consumption in the world. The sea's shoreline is now 100km (60 miles) from the old seaside promenades and, for most of the time, the sea itself is split into three lakes. Kazakhstan wants to protect one of the lakes, in the top of this picture, by damming its exit. But that will only hasten the disappearance of the other two. Meanwhile, without the moderating influence of the sea the climate of Central Asia is changing. Dust storms rage, poisoning local populations with salt and chemicals picked up from the exposed sea bed.

Kazakhstan

Uzbekistan

MALDIVES Indian Ocean

1995 The Maldives is a nation of 1,200 coral islands in the Indian Ocean, southwest of India. Many of the islands are unnamed and uninhabited specks of land. Nowhere on this particular island, in the centre of the archipelago near the capital Male, is more than 100m (300ft) from the shore, nor more than a few tens of centimetres above sea level. But such islands are far from worthless. Tourism is the country's biggest industry and visitors will spend large amounts of money to find their own undisturbed tropical paradise for a day, and to snorkel among the coral.

Rising tides Sea levels are rising and the beaches on these islands are eroding fast. This is the same shore on the same island less than a decade later. The sand is largely gone, dead coral has been exposed, and the palm trees are starting to capsize into the ocean. Many of the islands of the Maldives are so low that they were briefly inundated by the Boxing Day tsunami in 2004. Few lives were lost, but the scenes provided a fearful foretaste of the islands' likely future. As sea levels rise further, most of them could be permanently underwater by the end of the century.

CARYSFORT REEF, FLORIDA KEYS USA

1975 Until the late 1970s Carysfort Reef on Key Largo was the largest and most luxuriant coral reef among the islands that make up Florida Keys. It was not exactly undisturbed by humans. The reef is named after a famous ship, the British ship *HMS Carysford,* which was wrecked there in 1770. No fewer than 63 ships came to grief here in the treacherous waters around the reef in the 1830s, before engineers sunk piles into the reef to install a lighthouse – the first reef lighthouse in the world. But it had large areas of pristine shallow reef containing distinctive corals such as Elkhorn and Staghorn. And divers loved its steep "cliff", which contained a huge variety of coral formations.

Dying reef Carysfort Reef is now part of the Florida Keys National Marine Sanctuary. But that hasn't saved it. The decline in the ecology of the reef since the 1970s is pictured here by the same marine biologist who took the first image. Shortly after the first picture was taken, the reef was hit by two coral diseases: White Plague and Black Band Disease. Meanwhile, sediment from pollution and other human activities fed a smothering growth of algae. And there has been increasing physical destruction of the coral by boats. Carysfort Reef is still a popular diving destination today, but it has lost 90 percent of its coral since the 1970s, and is probably in terminal decline.

SAKHALIN FOREST Russia

1989 Russia contains most of the world's largest forests. Here are the forests of Sakhalin, a large island in the far east of Russia between the mainland and Japan. Like much of this remote region, it has historically been largely empty of humans except for a few hunters, loggers, and fishing communities. This is lucky, for these forests are extremely susceptible to fires. Fires are a natural part of their ecological cycle and usually start with lightning strikes. But human activity often brings many more fires, and can trigger widespread forest destruction. The red patches in this 1989 image show where forest has recently been destroyed by fire. All three burned areas are close to the main road down the spine of the island, suggesting they were man-made.

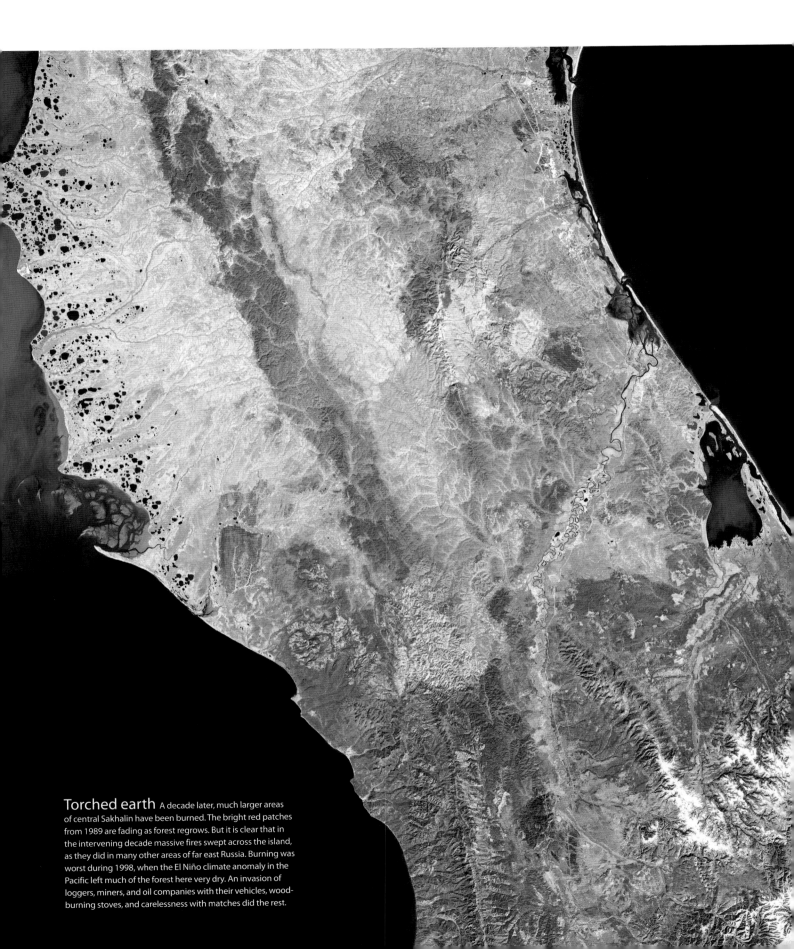

Torched earth A decade later, much larger areas of central Sakhalin have been burned. The bright red patches from 1989 are fading as forest regrows. But it is clear that in the intervening decade massive fires swept across the island, as they did in many other areas of far east Russia. Burning was worst during 1998, when the El Niño climate anomaly in the Pacific left much of the forest here very dry. An invasion of loggers, miners, and oil companies with their vehicles, wood-burning stoves, and carelessness with matches did the rest.

TIANANMEN SQUARE, BEIJING China

1997 Tiananmen Square, in front of the ancient Forbidden City in Beijing, is one of the greatest public squares on Earth. Its 40 hectares (100 acres) are famous for many things. It was here that students called for democracy in 1919 and again, when they were brutally crushed, in 1989. The Cultural Revolution began when a million young Red Guards rallied here in 1966, and the Revolution's instigator, Chairman Mao, lies here in a marble mausoleum. On a bright spring morning the square seems to sum up the best and worst about China and its capital city.

Sandstorms China has growing ecological problems, and they, too, are visited on Tiananmen. The country's 1.3 billion people and their breakneck dash for economic growth are placing huge strains on the environment. Fertile fields and pastures thousands of kilometres from Beijing are turning to dust as the country tries to feed its people. And, when the wind is in the right direction, those crumbling fields make their presence felt in the capital as huge sandstorms rage out of the west. This is Tiananmen Square in 2003, with the Forbidden City shrouded in sand. The storm will head out across Korea and Japan, cooling the Earth's surface and dosing marine ecosystems with minerals before giving a sandy dusting to the mountains of Canada.

OZONE LAYER Antarctica

2006 The ozone layer surrounds the entire planet in the stratosphere. It protects life on Earth by blocking out harmful ultraviolet rays from the sun. Without it there would be many more cancers, cataracts, and other diseases, and ocean life could be destroyed. But since the late 1970s a "hole" has opened up in the ozone layer above Antarctica each southern spring. The hole is caused by very stable man-made chemicals that slowly rise up into the stratosphere, where they release chlorine and bromine. During the long, dark, and uniquely cold winters over Antarctica, a "soup" of ozone-destroying chemicals builds up. As soon as the sun hits the atmosphere again, it triggers a runaway chemical reaction in which small amounts of the gases can destroy large amounts of ozone. In this image, from September 2006, the purple area shows where more than 75 percent of the ozone has been destroyed – and at certain elevations the figure is more than 90 percent.

The hole can heal This is the ozone layer three months later. It has largely healed. As the spring sun gradually warms the air, the runaway destruction of ozone stops and other reactions start to reform more ozone – until the next year. Global controls on ozone-destroying chemicals such as CFCs mean that their concentrations have now stabilized in the atmosphere. Soon they should start declining. When that happens the annual Antarctic ozone hole should gradually diminish. In perhaps 50 years, the ozone layer may resume normal service and this picture could be typical of September, too.

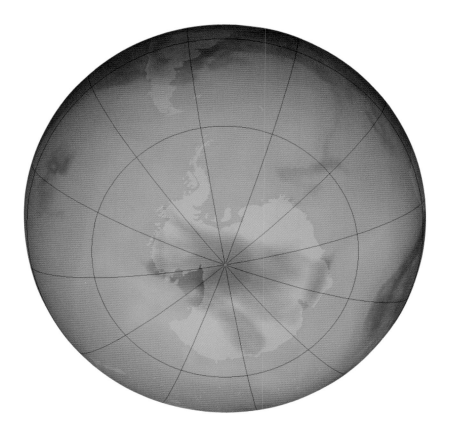

Urbanization

This decade, for the first time in history, most humans live in cities. Urban areas still occupy only two percent of the Earth's land surface, but they house more than half of us. The unprecedented migrations of people from the countryside mean that virtually all of the world's population growth is now in cities. Around 1940 New York became the world's first megacity with a population of more than 10 million. Today there are 20 of these urban monsters; Tokyo now has 27 million people; São Paulo and Mexico City are not far behind.

Great cities and ancient trading ports such as Beijing, Dubai, and Singapore have grown, but others, such as São Paulo, have emerged from obscurity to dominate their countries. Still others, including Las Vegas in the desert of the American West, have been created in just a few decades from blank places on the map.

Cities are the engines of economic growth. They grow around activities best carried out centrally, such as government and finance, manufacturing, wholesaling, and running ports. They become centres of skilled labour and sure markets for goods, so that even when their old industries subside, new ones are able to take their place.

Cities often grow despite being noxious places, centres of disease and pollution. Tens of millions of Chinese peasants flock to cities every year, even though half a million Chinese die each year from lung disease associated with urban air pollution. See here the visual evidence of how Kathmandu, once a Himalayan idyll, is now choking on traffic fumes accumulating in the thin mountain air. And cities can be violent. The biggest cause of death among young people in São Paulo is homicide, followed by traffic accidents.

But, let's face it, people like living in cities. The streets may not be paved with gold, as Dick Whittington had hoped when he headed for medieval London, but at least they are paved. Even for the poor, cities represent the good life and the bright lights, a fact graphically illustrated by our pictures of Los Angeles. But cities also have a tremendous capacity to reinvent and improve themselves. Look at the transformation of Bilbao from an old Basque centre of heavy industry to a beacon of 21st-century culture; or the reinvention of the river through Seoul; or the contrasting regeneration of rundown Buenos Aires and the British military port of Gibraltar. See, too, how the old West India Dock in London, once the world's greatest imperial commodities port, has become Canary Wharf, one of the great global centres for trading in money and information, and how Seattle has gone from smokestacks and shanty towns to Microsoft and *Frasier*.

The products (and pollution) from cities extend far beyond their borders. But so, too, do their demands. Cities consume three-quarters of the resources we take from the Earth. Maintaining London is reckoned to require a land area 120 times that of the city itself. So look elsewhere in this book for evidence of how the demands of the world's cities impact on the rainforests of Brazil, the wheat prairies of North America, the cotton fields of central Asia, the copper mines of Chile, and, ultimately, on the ice sheets of the polar regions. The urban footprint in the 21st century is truly global.

São Paulo has gone from being a Brazilian backwater to one of the world's three largest cities in little more than a generation. And nowhere better reveals the gulf between the haves and have-nots in our new urban world. The beneficiaries of the city's boom live and work in its forests of high-rise blocks – such as this one in the wealthy suburb of Morumbi. But, in their shadow, many more citizens cling to the underside of the urban revolution. They are the 60,000 residents who live in the *favela* of Paraisopolis.

HONG KONG China

1880 The barren Chinese island of Hong Kong was a neglected fishing port with a population of just 7,000 until it was taken over by the British in 1841 after the Opium Wars with China. By 1880 the sheltered waters between the island and the mainland had become a major port, a bustling centre for east-west trade, and a centre for banking and insurance for the British Empire throughout Asia. Docks, hotels, and warehouses dominated the island's narrow coastal strip, seen in the foreground, while ferries crossed constantly to the Kowloon Peninsula on the mainland, seen here in the distance. By 1880 Hong Kong's population was approaching 160,000.

Rapid expansion Hong Kong reverted to Chinese control in 1997. It has long since outgrown its waterside origins, spreading upward into the mountainous centre of the island, as well as outward across Kowloon and the New Territories beyond. Its population of more than seven million now occupies one of the most densely packed conurbations in the world, with more than 6,000 people on every square kilometre (0.4sq mile). High-rise buildings dominate the island's coastal strip, the precipitous lower slopes of the central mountain, and the artificial land created by developers at the water's edge. The tallest building, pushing above the skyline at the centre-left of the photo, is the 88-floor International Finance Centre. Hong Kong and the neighbouring Chinese province Guandong now seem set to merge, forming the largest urban agglomeration on the planet.

ABERDEEN HARBOUR, HONG KONG China

1930s The south side of Hong Kong Island contains a distinct town known as Aberdeen, after the Earl of Aberdeen, a British war minister in the mid-19th century. Its harbour, sheltered by other islands such as neighbouring Ap Lei Chau, has for centuries been full of junks and houseboats. Home to the Tanka boat people, it was also a haven for the pirates of the South China Sea. And its floating restaurants were famous in the region even before the arrival of the British on the island.

Floating economy In the past half century Aberdeen, like the other side of Hong Kong Island, has embraced modernity. Its engineers have built highways on land reclaimed from the harbour, and high-rise apartment blocks up the hillsides. They have even drawn up plans to fill in the harbour to make more land for development. And yet, for now, the small boats remain. Aberdeen harbour still has numerous boat yards and a huge fish market; fish restaurants line the shore and fish are reared on floating farms out in the harbour, ready to supply the fast-growing market for live fish in restaurants both in Hong Kong and on the Chinese mainland.

DUBAI United Arab Emirates

1940s As recently as the 1960s, Dubai was a traditional Arab trading city on the Persian Gulf. Its chief commodity back then was gold. More than 200 tonnes of the stuff passed each year through Dubai's numerous marketplaces, most of it smuggled into India. Its central Creek was still lined with the low-rise homes and business premises of its traditional leaders and traders – often complete with wind towers, a natural form of air conditioning.

Built on trade Oil was discovered near Dubai in 1966. But Dubai remains more of a trading city than any of its partners in the United Arab Emirates. It is an *entrepot* for the globalization of trade in everything from plastic waste to the latest electronics. It has one of the world's largest artificial ports. And much of the profit is being invested in transforming the city's skyline, particularly along the Creek. Dubai is today one of the fastest-growing cities on Earth, with an army of a quarter-million mainly Indian and Pakistani workers engaged in construction projects with an estimated value on completion of $100 billion. One thing is for sure: all the new buildings will have modern air conditioning.

MEXICO CITY Mexico

1986 With a population of 20 million, Mexico City is one of the world's largest cities. It also used to be the smoggiest city in the world. The surrounding mountains trapped industrial pollution and traffic fumes from ill-maintained buses, taxis, and cars. This is also one of the world's highest major cities, and the thin air at 3,000m (9,800ft) slows the atmosphere's natural cleansing processes. As a result, for much of the year the mountains were barely visible. Legend had it that breathing the foul air for a day was as dangerous as smoking a packet of cigarettes. The city's reputation became so bad that international corporations were pulling out for fear of damaging the health of their executives.

Cleaner air Since the late 1980s the city authorities have worked hard to clean up the smogs. They banned high-sulphur oil from factories and cars, imposed checks on vehicles every six months to ensure clean running, and restricted traffic during the heaviest pollution episodes. Nobody would say the smogs are banished, but some days, as here, the mountains can be seen clearly. And, in a city with ever more people and cars, at least things haven't got worse.

SAN FRANCISCO USA

1906 At the start of the 20th century San Francisco was America's smartest, newest, and most go-getting city. With some 400,000 inhabitants, it was home to a quarter of the fast-expanding population of Americans west of the Rocky Mountains. But just before dawn on 18 April 1906 the northern end of California's now-notorious San Andreas Fault shifted. In the resulting earthquake the city's buildings tumbled and fires spread through the wooden buildings. The fires did most damage, devastating most of the city – as this image looking down California Street a month later shows. The Grace Church stands out amid the ruins. The earthquake killed some 3,000 people and left most of the city's population homeless.

Rebuilding programme San Francisco, of course, was built afresh. It was laid out in a grand manner, with wide arterial roads through the once-ramshackle city centre and a subway system. Reconstruction was largely completed within a decade. The buildings in the foreground in today's photo, taken from the same spot as the 1906 image, come mostly from this phase. But in the past 50 years a whole new central business district of high-rise blocks has risen downtown. Today, the Ritz Carlton Hotel stands on the site of the former Grace Church. The church itself was eventually rebuilt as Grace Cathedral in the suburb of Nob Hill.

CITY HALL, SAN FRANCISCO USA

1906 By the time the earthquake had finished rocking San Francisco on 18 April 1906, and the fires had been put out, the City Hall was one of the few buildings to be recognizable. But the fires had destroyed most of its contents, including the city's archives. One of the unexpected consequences was that tens of thousands of temporary Chinese migrants who had colonized the city were able to claim permanent residency, arguing that the Hall had contained records of their citizenship. They soon brought in their relatives, boosting the city's Chinatown – one of its most significant modern cultural features.

New City Hall After the earthquake the city fathers decided to tear down the remains of the old City Hall, sturdy though it had proved during the quake. They replaced it with another – bigger and bolder, but bearing a strong similarity to its predecessor. No expense was spared. The new City Hall was completed in 1915 at the then-extraordinary cost of $7 million. The building occupied 46,000sq m (500,000sq ft), its dome was among the world's largest, and behind its facade of granite, marble, and sandstone it contained 8,000 tonnes of steel shipped from Pittsburgh. But it was not earthquake-proof: the mighty dome was twisted on its base by 10cm (4in) during a small quake in 1989.

SEATTLE USA

1937 The 1930s was the decade when the American dream almost fell apart. Millions of people attracted to the cities by jobs found themselves out of work during the Great Depression. Unable to pay rent, many were thrown out of their homes as well. In many cities they took refuge in self-built hovels on spare land, such as here at Hooverville close to the port of Seattle. The people here had built the downtown art deco towers in the distance, and queued for jobs in the city's belching factories. But now they were unwanted and virtually under siege. City authorities several times tried to tear down Hooverville, which became a national symbol of resistance to unbridled capitalism.

Slicker city Today Hooverville is long gone, replaced by a giant container yard for the port. The descendants of its inhabitants now mostly have jobs, and homes in the suburbs. Heavy industry, too, has been banished, and the city air is much cleaner as a result. Seattle is now the city of Boeing, Microsoft, and Frasier, and its downtown high-rise has grown ever higher. But some 1930s buildings remain: notably the tall building from the top right of the first picture, which now appears as a midget in the shadow of taller newcomers.

SANTA CRUZ Bolivia

1975 Santa Cruz de la Serra, in the bottom left, is Bolivia's second-largest city. It sits on the western edge of a vast area of lowland forest, bush, and wetland that once stretched almost uninterrupted to the Atlantic Ocean, 2,000km (1,200 miles) to the east. As can be seen here, as late as the mid-1970s the forest east of the River Grande was largely empty. But around then, the government decided to move tens of thousands of people from the poor highland regions of the Andes to clear the forested lowlands and plant crops in the fertile forest soils.

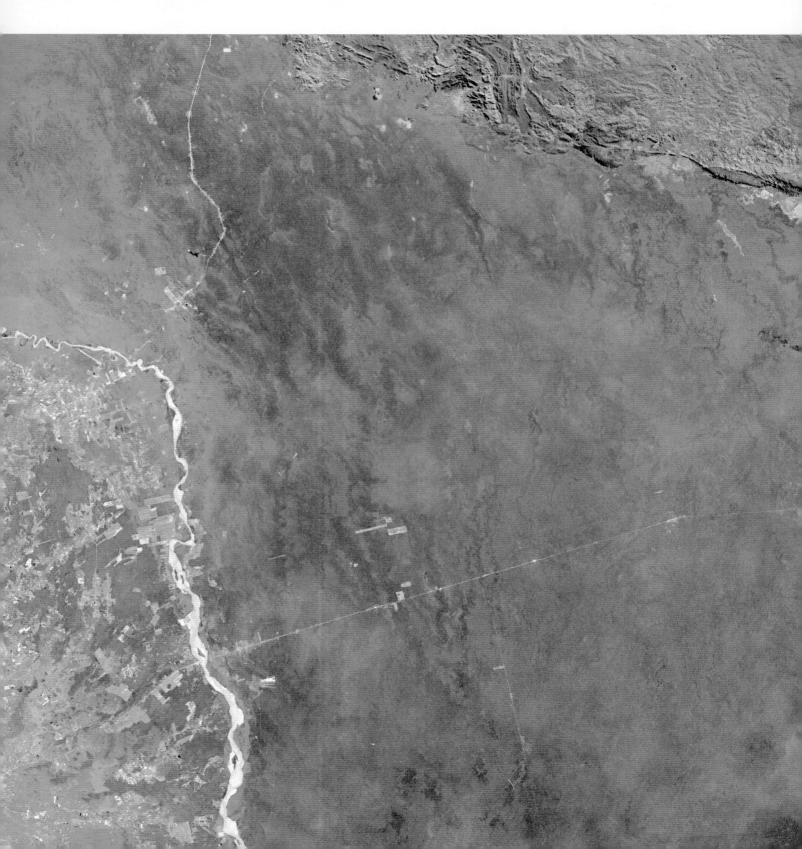

Urban growth Today the forest is gone – clean-cut to make way for a huge development scheme called Tierras Baja. That scheme now covers most of the area and is connected to the outside world through a new network of roads. Migrants live in village communities, each with a church, school, bar, and soccer field. The rectangular light-coloured areas seen here are fields of soybeans, which is the most valuable and widespread crop in the region. Most soybeans are exported to Europe and North America.

TOKYO Japan

1925 In the Land of the Rising Sun, Ginza always rises again. For more than a century it has been Tokyo's premier shopping district. It was first wrecked in 1872 when a massive fire ripped through its wooden buildings, after which the government hired an English architect called Thomas Waters to rebuild it in Western style, this time with bricks. Then in 1923 came the great Kanto Earthquake, which levelled almost every building. But this picture shows how, just two years later, Ginza was well on the road to recovery. Yet larger buildings had emerged, including the famous Waco department store at centre-left, and the street is bumper-to-bumper with trams.

Rising power Ginza lost its buzz during the austerity of World War II. And its buildings were destroyed again one night in early 1945 when Allied bombing created a firestorm through Tokyo that killed an estimated 100,000 people – more, even, than the atomic bomb at Nagasaki. But, undaunted, Ginza rose again as a brightly lit consumer paradise, the epitome of modern Japan and home of every electronic gadget imaginable – and the $10 coffee. This time it is built of steel and glass. But will it prove more durable?

BUENOS AIRES Argentina

1985 Buenos Aires, the capital of Argentina, has long been one of Latin America's richest and most Europeanized cities, with plenty of upmarket neighbourhoods that would not disgrace Madrid or Milan. But there has always been another side to the city: its poor and crime-ridden shanty towns, or *barrios*, which have few counterparts in Europe. One of the most notorious has been behind the railway station in the otherwise-chic inner-city suburb of Retiro. Here, within sight of the chrome and glass of modern Buenos Aires, live the city's underclass – the waiters, dog-walkers, and production-line workers who serve the rich of the city.

Self-help scheme It was only in the 1990s that paved streets and basic services such as regular electricity and clean water arrived behind the railway station. But with the arrival of those services the *barrios*' residents have begun to construct a better community. However, self-help is the order of the day. Twenty years after the previous photo was taken, photographer Mark Edwards returned to the same spot and found that the new residents were replacing the thin plywood and corrugated iron shack with a new home of bricks and concrete. Retiro, at long last, is on the up.

SKYE BRIDGE Scotland

1995 One of Scotland's best-known folk songs, *The Skye Boat Song*, describes the escape of Bonny Prince Charlie "over the sea to Skye" after his defeat at Culloden in 1746, the last major battle fought on the British mainland. The romance lingered, and Skye, one of the largest of the islands of the Inner Hebrides, long cherished its separateness. In modern times the railway and road both came as far as Kyle of Lochalsh, the mainland town on the right-hand shore here. After that, would-be visitors to Skye took the short ferry ride over Loch Alsh, much as Charlie must have done.

Bridging the gap Everything changed in October 1995 with the completion of the Skye road bridge. The ferry shut and the romance of the separateness of Skye was lost. It was, strictly speaking, no longer an island. There was another problem. Much as the locals had hated the constant queues for the ferry, they hated the high toll charges on the bridge even more. They made constant complaints until the toll was abolished in 2004.

MILLAU VIADUCT France

2001 The Tarn Gorge had long been a traffic bottleneck for drivers on the *Route Nationale* N9 near Millau in southwest France. It is one of the main routes from Paris to Languedoc and Spain. Every summer, traffic built up as the road zigzagged down into the 200m- (650ft-) deep gorge, across a small bridge over the river, and back up again. The gorge was such an environmental blackspot that even green activists were generally in favour when the government announced plans for a bridge to span the valley.

Seven-pier wonder After four years' construction the Millau Viaduct opened in December 2004. It stretches for 2.5km (1.5 miles) across the top of the valley and is the tallest road bridge in the world. One of its seven piers soars higher above the valley bottom than the Eiffel Tower does above Paris. Designed by British architect Sir Norman Foster, the bridge has been hailed as a design marvel, and a platform for viewing the bridge has become a major tourist attraction in its own right.

SPROGO ISLAND Denmark

1988 Sprogo is a small island in the middle of the Storebaelt, or Great Belt – one of two major natural waterways linking the Baltic Sea to the North Sea. It has a lighthouse, a cottage for the lighthouse keeper, a jetty, some rudimentary sea walls, a few trees, some intriguing Neolithic remains, and that's about it. On the map it looks like a stepping stone between the larger Danish islands of Funen and Zealand, the latter of which houses the Danish capital, Copenhagen. And that's how things have turned out.

Extended span In the 1990s engineers constructed the 17km (9-mile) Storebaelt link between Zealand and Funen, via Sprogo. The western bridge in the distance, between Funen and Sprogo, has two decks carrying a motorway and a railway. Stretching east towards the camera on the left is the start of a 7km (4-mile) suspension road bridge from Sprogo to Zealand. And on the right is the start of a railway tunnel to Zealand. In the process of creating this mid-ocean transport hub, land reclamation has quadrupled Sprogo's size. The project forms part of a grander scheme for a series of giant transport links to span several entrances to the Baltic Sea, providing a direct link to northwest Europe for the first time.

VALE OF NEATH Wales

1994 The valley of the River Neath was one of south Wales' most peaceful, with woodlands, fields, a handful of small villages, and the Glyncorrwg hills behind. But transport engineers had had their eyes on the valley for a long time. It was earmarked to form a vital link in the planned A465 "heads of valleys" route through the former mining regions, from Abergavenny to Swansea, and become a new part of Europe's trunk-road network. And, in 1984, the planners got their way.

Major road link This is the same scene three years later. The £80-million Aberdulais to Glynneath link road is complete. Fields have been turned to concrete and the rural tranquillity is gone for good. The valley, once most famous for its walking trails to the Aberdulais Falls, is now zoned for economic development. Some want to see industry return to a valley that used to be home to miners; others foresee a new "industrial heritage industry" based on the Neath Canal.

BEIJING China

1978 In the last days of China's revolutionary communist leader Mao Ze Dong, Beijing was a relatively austere and tightly packed capital. It forms the small grey area in this satellite image. The hills to the west are still green with forests, and farms show as red, orange, or yellow, depending on the crop being grown. Chairman Mao's era had lionized the rural peasant and there were strict controls on people moving to the cities, where, in any case, all jobs were under state control. Economically, the capital of the world's largest country, already with around a billion inhabitants, slumbered.

City migrants Economic reforms begun in 1979, three years after Mao's death, unleashed private capitalism on China, albeit under state control. The country began more than two decades of rapid economic growth, with GDP increasing by about 10 percent a year. China's population growth was capped by a rigorous birth control policy, but as the economy boomed, millions of migrants flooded from the countryside into Beijing and other cities. This second image shows how the city's suburbs have grown to house the migrants, taking over huge swathes of former farmland and engulfing market towns. The city's population has almost doubled in size to 15 million.

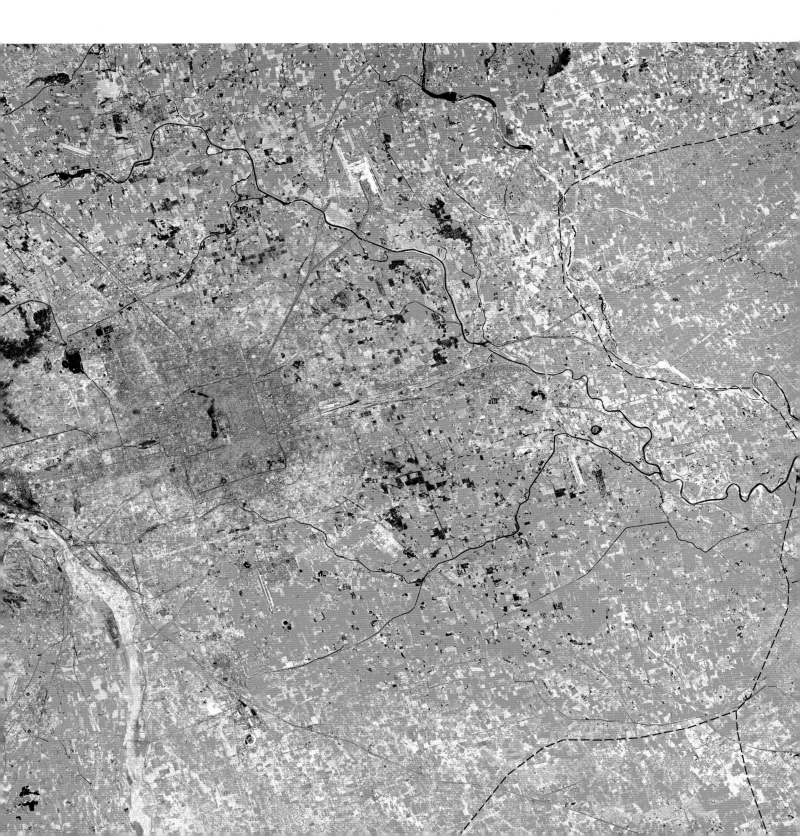

AVENUE PAULISTA, SÃO PAULO Brazil

1902 In the late 19th century São Paulo was a Brazilian backwater, smaller and less famous than Rio de Janeiro down the coast. Avenue Paulista was one of its major thoroughfares. By the turn of the century, when this picture of the avenue was taken, the state's coffee industry was booming, bringing in millions of migrants from Europe to work the plantations. Many of the estate owners lived on the avenue, which remained a relaxed, spacious, and tree-lined retreat from the bustle of global trade.

Megacity Avenue Paulista today is no retreat – it is the hub of one of the world's great megacities. With a population of some 19 million, São Paulo's metropolitan area ranks only behind Tokyo and Mexico City. The city's coffee business, though still vibrant, has been overtaken by many others, from steel to car manufacturing and textiles to electronics. And many of the finance houses and corporate headquarters that are making Brazil's economic growth second only to that of China are on Avenue Paulista. Not surprisingly, it has some of the most expensive real estate in Latin America. But down there, somewhere, a few of the original mansion houses remain as reminders of a more leisurely past.

Nepal **KATHMANDU**

1970 High in the Himalayas, Nepal is a land of goats and small fields, of clear air and rural communities connected only by long footpaths through the mountains. In 1970 virtually the entire population of Nepal lived in the countryside, much of it extremely remote. This is a scene 1,300m (4,300ft) up in the Kathmandu valley, several kilometres from the nation's small administrative capital of the same name.

Urban sprawl The lower shot shows the same spot today, and was taken by the same photographer. The houses, the village, and the fields are all gone – the sprawling suburbs of Kathmandu have consumed them all. A city of only 200,000 people in 1970, having engulfed its neighbours Patan and Bhaktapur and most of the land in between, it now has a population of more than 1.5 million. Unlike in many other Asian countries, urbanization in Nepal has not been accompanied by rising living standards. The country remains one of the poorest, least industrialized in Asia. People have come to Kathmandu to escape grinding poverty and in the hope of finding basic services such as education and health care. But at this spot, villagers have become city-dwellers without moving. Of the former landscape only the mountains remain – though they are so shrouded in the smog accumulating in the rarefied air that they are barely visible.

SYDNEY Australia

1975 Australia, the sixth-largest country in the world, is also one of the least densely populated. An area almost the size of the USA has just 19 million people. But most of them live in a thin strip of land between the east coast and the Great Dividing Range. The largest city, Sydney, had a little under three million people in 1975, when this satellite image of the city and its surroundings was taken. The light areas are already urban, the predominantly light green areas are bush, and the darker green parts are forested mountain and national park.

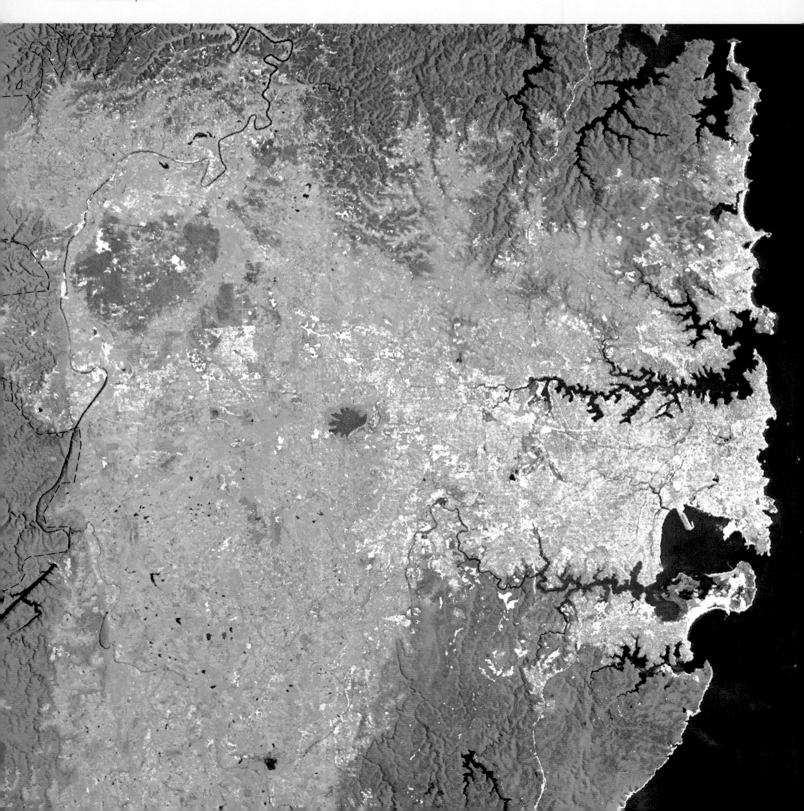

Invading the bush
In the three decades since, Sydney has grown by almost 1.5 million people. Most have been housed in sprawling suburbs to the west, engulfing towns such as Camden and Richmond, and resulting in huge areas of bush being paved over. The city now extends right up to the River Nepean, beyond which lie the Blue Mountains, part of the Great Dividing Range. This extension has been bad for the natural environment, but dangerous too for the residents of these outer suburbs, who are ever more vulnerable to bush fires each summer.

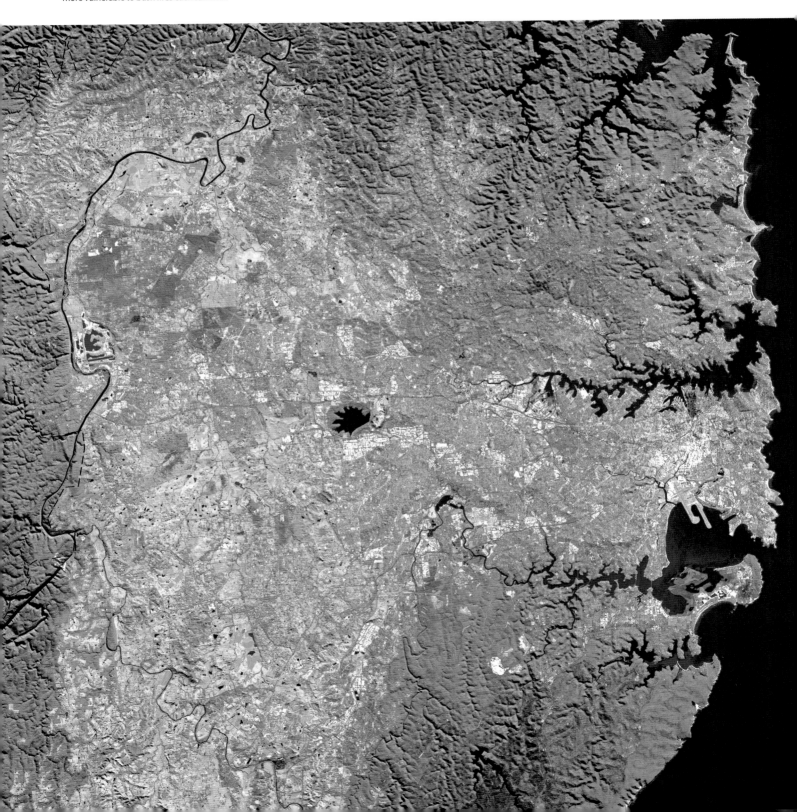

GIBRALTAR Mediterranean Sea

1954 The Rock of Gibraltar occupies the southernmost tip of the Iberian peninsula. Though claimed by Spain, its big neighbour, the 6.5sq km (2.5sq miles) of Gibraltar have been a British dependency since the Treaty of Utrecht in 1713. The harbour, which commands the entrance from the Atlantic Ocean to the Mediterranean Sea and the Suez Canal, has been occupied continuously by the Royal Navy. Though lacking the grandeur of Asian city-states such as Singapore and Hong Kong, Gibraltar has tried to lead an independent existence – catching most of its water from tanks set into the Rock, for instance. Half a century ago, however, it remained little more than a naval outpost, with meagre housing creeping up the slopes of the Rock.

Steady growth Spain continues to block access to the airport for flights other than to Britain, but otherwise relations have improved now that both Spain and Britain are in the European Union. The dockyard accounts for only seven percent of the dependency's economic activity, compared with 60 percent as recently as 1984. And modest prosperity has come to the Rock, albeit sometimes due to its convenience as a tax haven, a centre for online gambling, and as a staging post for drug-running from North Africa into Europe. Certainly the marina is flourishing, and high-rise buildings have grown along the shore.

CANARY WHARF, LONDON England

1976 London's West India Docks were among the world's largest when they opened in 1802 – unloading produce such as rum, sugar, and molasses shipped from British colonies in the West Indies. The docks occupied most of the Isle of Dogs and were connected to the River Thames by giant locks. They could berth 600 ships at a time, decanting their contents into more than a kilometre (0.6 miles) of brick warehouses. But German bombing during World War II destroyed most of the warehouses, and by the time this picture was taken most of the business had departed for bigger ports, such as Felixstowe, that could handle containers. The dock finally closed in 1980.

Urban regeneration

After being derelict for less than a decade, the Isle of Dogs was reborn. The old hub of Victorian imperial trade has been replaced by an axis of 21st-century global finance. Canary Wharf, named after the part of the West India Docks that handled trade to the Canary Islands, is the heart of the regeneration of London's Docklands. It contains Britain's three tallest buildings, employs more than 80,000 people, and houses major banks, law firms, and two national newspapers. Already a rival to London's traditional financial centre, "the City", 6km (3.7 miles) upstream, there are plans to double its size. Behind it, on the opposite bank of the Thames, is the distinctive white shape of the Millennium Dome – a less successful symbol of modern London.

SEOUL South Korea

2003 Through the 1960s and 1970s Seoul, the South Korean capital, was one of the fastest-growing cities on Earth. Effluent discharges reduced its main river, the Cheonggyecheon, to little more than a sewer. So the government completed the job by encasing the river in concrete and building a six-lane highway above it. From London to Tokyo, many urban rivers have been killed off in this way – a final act to banish nature from the urban environment.

River revival But in Seoul the mayor wanted to turn the tide. In 2002 he decided to get rid of the road, restore the river, and create an 8km (5-mile) park along its banks. Despite widespread opposition, the restoration has been achieved. Drivers were offered an improved bus service. Besides improving the city's landscape and providing a centre for recreation and wildlife, the revived river also acts as an air conditioner for the city. On its banks summer temperatures are more than 3°C (5.4°F) lower than when it was concrete. Other cities, including Shanghai and Los Angeles, are now following Seoul's example.

DOWNTOWN Singapore

1962 Singapore has for more than a century been one of the great trading ports of the world. In 1962, in the midst of a messy divorce from Malaysia, the 50km- (30-mile-) long island state housed some 1.6 million people. Three stages in its development are shown here: the rowing boats and sampans of its Malay Chinese nautical origins; the old colonials' shorefront buildings from the height of the British Empire; and the 1950s building to the left, the Asian Insurance Building, which was at the time the city's tallest.

Upward mobility Since breaking with Malaysia, independent Singapore has grown rich as a port and financial centre to rival London, New York, and Tokyo. It became the smallest but most dynamic of the "tiger economies" of Asia. Since the 1960s it has grown vertically to accommodate a population that now numbers 4.5 million.

Most remarkably, notice the Asian Insurance Building in this modern picture. Nestled close to the far left of the picture, the building that was once the city's largest is now the smallest thing in the central business district.

BILBAO Spain

1948 Bilbao has long been one of the "smokestack" cities of Spain. Feeding on the iron reserves in nearby hills, it became a centre of the steel and shipbuilding industries in the heart of the Basque country, and for a long time the richest city in Spain. But wealth came at a cost. Fire and smoke often filled the air, as in this dramatic post-war industrial landscape in the dock area of Euskalduna and close to the Basconia steelyard.

Cultural shift Almost half a century on, Euskalduna has been transformed into a centre for services and culture. Like many centres of heavy industry in Europe and North America, it shed its former economy activities almost entirely in the 1980s. With the shipyards and steel mills gone, Euskalduna is now a waterfront development of parks, apartments, shops, and offices. And this totemic piece of architecture – the Guggenheim Museum, which opened in 1997 and is dedicated to 20th-century art – symbolizes the change. Only the ship shape seems vaguely familiar.

LAS VEGAS USA

1973 "Vegas" was founded in 1905 as a desert railway town used by the silver and gold miners of Nevada. For much of the time since, it has been the fastest-growing urban area in the USA. First, it was a base for construction on the nearby River Colorado of the Hoover Dam, the world's first superdam. And since the 1940s it has developed a new identity as a centre for hotel-based casinos, entertainment, and conventions, fuelled partly by the nationwide move from the north and east to the "sunbelt" states of the south and west. At the time of this image the permanent population was around a quarter-million.

Watering the strip Today, driven by the development of resorts such as the Mirage, which opened in 1989, the city has spread across most of the available desert floor. It is approaching the Spring Mountains to the west. The population of the urban area exceeds a million, and the growth of the paved area across the desert floor has accentuated the danger from flash floods coming out of the mountains during the occasional heavy rains. And it suffers, too, from water shortages as local aquifers are pumped out to maintain the city's fountains and water its parks and palm trees.

LOS ANGELES USA

1908 Los Angeles began as a farming village on the banks of the Los Angeles River in the late 18th century. By 1908, when this picture was taken from the summit of Kitt Peak 30km (19 miles) away, it was growing fast. It had doubled its population to more than 250,000 since the start of the decade. Its street lights were still just a pool of light in one corner of the Los Angeles Basin, but the city had big plans. Film studios had just opened for business in Hollywood, and the city fathers were building a reservoir and aqueduct to tap water from 300km (200 miles) away in Owens Valley, which would underpin the future expansion of the desert city.

Light pollution Now a megacity of some 13 million people, Los Angeles and its suburbs burn bright right across the valley to the base of Kitt Peak. The city is visible for hundreds of kilometres. The searchlight on the top of one hotel can be seen some nights in Death Valley 150km (100 miles) away. But all that light consumes a lot of energy in climate change-conscious California. And the glare means that stars are no longer visible in the night sky. LA is today trying to cut its light pollution, but clearly there is a long way to go.

Land Transformation

Mankind has been exploiting the land ever since we left our first footprints on the Earth. For thousands of years we have ploughed pastures and chopped down forests, gouged minerals from the Earth and irrigated deserts. But it is only in the past two centuries or so, since the beginning of the industrial revolution, that our imprint on the planet has become truly indelible. And the great majority of our impacts have been within that relatively minuscule time frame that is the past half-century.

Take mining. Winning metals, minerals, and fuel from the Earth is the world's largest industry: uranium for bombs and phosphates for fertilizer; aggregate for roads and copper for cable; oil to fuel your car and coal to generate your electricity; polyester for your shirt and coltan for the capacitor in your mobile phone; bricks for your house and stones for your patio; talc for your talcum powder and diamonds for your jewellery; aluminium for aircraft and steel for almost everything. All come from the Earth.

Today the easy supplies of most of these vital resources have run out. Ever-larger amounts of rock ore have to be extracted to keep industry supplied. Production of copper has increased 22-fold in the 20th century. But the amount of ore ripped from the Earth to produce that copper has increased more than a hundredfold. Hence the huge holes in the ground where copper is mined at Bingham Canyon and Escondida. Other metals have the same problem. Witness the vast strip mines removing bauxite, the raw material for aluminium, around Weipa in Australia. Thanks to recycling, most of the world's aluminium is reused several times but, even so, the mines keep growing.

Mining has laid waste huge areas of forest and farmland around the world, and forced the expulsion of an estimated 100 million people from their homes. On Nauru, miners have turned an entire Pacific island into a moonscape to provide nitrate fertilizer for Australian farmers. Some of the world's former mines, as we show, have been partially rehabilitated. That is good news. But bringing nature back is often hard, and has not even been attempted in most places. The scars will often still be there in millions of years.

The products of mining create pollution that fills the air in cities and eats away at great monuments. Meanwhile, forestry and farming, too, now take place on an industrial scale. This chapter shows the evidence of how some of the largest and most remote forests have been logged and their land turned into plantations or raising grounds for prawns.

In addition to these direct impacts on the terrain, huge dams now plug great rivers such as the Euphrates in Turkey, the Colorado in the USA, and the Yangtze in China. They do so in order for us to generate electricity and provide water for irrigation projects that grow cotton, wheat, and even rice in the world's dry lands. The flow of these and many other rivers is now under total human control. And we do not always use that control wisely. The Aral Sea in central Asia, once the world's fourth-largest inland sea, has been reduced to three saline sumps surrounded by a vast toxic desert.

Timber is one of the world's most important commodities, used in construction, to make paper, and in the manufacture of furniture and much else. Most of the world's forests today are little more than timber mines. And none more so than here in southern Finland, close to the Russian border, where Lake Saimaa provides a convenient collecting point for the region's primary resource. Huge rafts of logs head for a string of giant sawmills on the southern shore of Finland's largest lake.

GLEN CANYON USA

1958 The Glen Canyon, an extension of the Grand Canyon on the River Colorado, was one of America's finest natural features. But for decades US engineers eyed the Colorado as a source of both hydroelectricity and water for irrigating the arid American West. They built the Hoover Dam downstream in Boulder Canyon in the 1930s, and by the 1950s they were keen to build another.

Damming evidence And they did. This is the Glen Canyon today. Plugged by a dam in 1964, the canyon was renamed Lake Powell – in memory of John Wesley Powell, a one-armed Civil War hero who, in 1869, made the first boat journey by a white man through the canyon. The reservoir took 17 years to fill. The Colorado River is, as a result of the Hoover and Glen Canyon dams, one of the most regulated rivers in the world. The two reservoirs can hold several years' flow, and virtually no water ever escapes from the river into the ocean. In recent years drought has lowered reservoir levels, revealing tantalizing glimpses of the old canyon – such as the white "bathtub ring" in this 2003 image. And, more than 40 years after the canyon was flooded, there is a growing campaign to break the dam, empty the reservoir, and bring back the canyon – permanently.

THREE GORGES China

1997 Since the 1930s, successive Chinese leaders have dreamed of damming the Yangtze, the world's third-largest river, where it exits a scenic mountain region known as the Three Gorges. They wanted to tame the great floods that sometimes bring havoc downriver, to make the Yangtze navigable all the way to Chongqing 1,300km (800 miles) inland, and to generate hydroelectricity that could industrialize the country's interior. Finally, in 1994, and despite growing opposition from environmentalists and people whose land would be flooded, the project got under way. Here, three years into construction, an endless line of trucks dumps earth and rock into the ever-narrowing gap across the river near the village of Sandouping.

Controversial barrier The dam is now complete. And the 600km-(400 mile-) long reservoir behind is filling. Three Gorges is set to be the world's most powerful hydroelectric dam. Its 26 turbines have a capacity of 18,000 megawatts. Its huge sea locks will let ships upstream to Chongqing, which is now one of the world's fastest-growing cities. But the costs have spiralled to an estimated $100 billion. Numerous archaeological sites have been flooded. And two million people have been removed from their fertile valley homes and relocated, often on poor hillsides.

ATATURK DAM Turkey

1982 The River Euphrates, which flows from top right to bottom left through this satellite image, has watered some of the world's earliest and most powerful civilizations on its journey from Turkey through Syria and Iraq to the Persian Gulf. Many of those civilizations had manipulated its flows, but nobody had attempted to halt the river's flow altogether. Not until the 1980s, when Turkey began to assemble a cascade of dams in the Anatolian highlands where the river rises and gains most of its water.

Water hoarding Turkey's Southeast Anatolia Project is intended to develop the eastern, Kurdish half of the country by using the river to generate hydroelectricity and irrigate crops. The centrepiece of the project is the Ataturk Dam, a 184m (600ft) wall of rock and earth across the river near Urfa. It was completed in 1990 at a cost of $3 billion. The huge reservoir, seen here, now covers more than 800sq km (300sq miles) and has displaced some 40,000 people. At times during the early 1990s, while it was being filled, the dam stopped all flows downstream into Syria and Iraq. Today, the dam's operation continues to upset the river's flood regime, helping dry out the Mesopotamian marshes in southern Iraq.

ANATOLIA Turkey

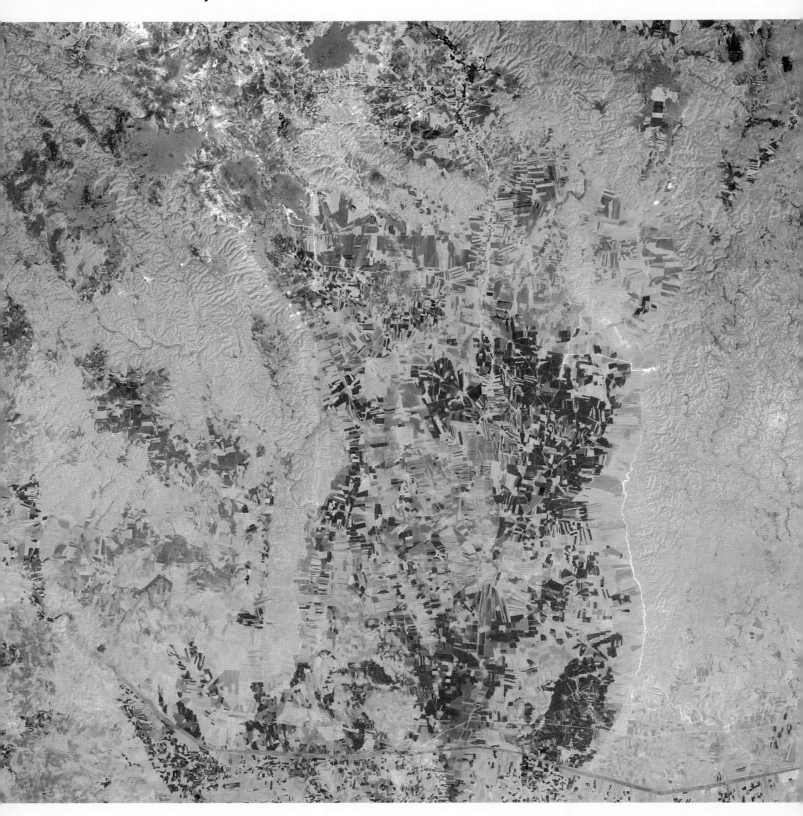

1993 Turkey's Ataturk Dam was built to capture the waters of the River Euphrates as they rush out of Anatolia in southeast Turkey towards Syria and Iraq. One aim was to irrigate the fertile valleys of Anatolia and bring prosperity to the Kurdish people of the region. This image shows the Harran plains, a large flat area between the reservoir and the Syrian border to the south. It was taken in 1993, shortly before the completion of two tunnels to take water from the Ataturk reservoir to the plains. The dark areas represent fields of cotton, a thirsty crop that farmers then irrigated with underground water reserves that they pumped from deep boreholes. But the underground water was running out.

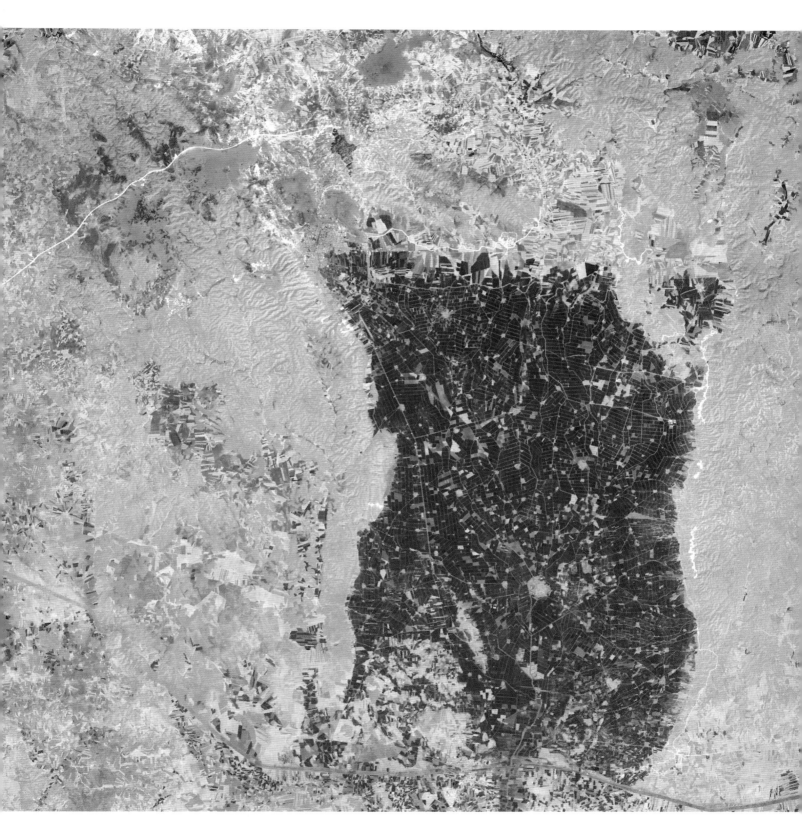

Selective irrigation A decade later the valley has filled with the dark green of growing cotton. Many of the boreholes are abandoned and farmers now rely on water delivered from the dam. The project has allowed an extra 200,000 hectares of land in the plain to be brought under cultivation. The contrast could hardly be greater between the cotton fields and the dry and barren land to the east and west in this image, taken in late August. What about downstream? Nobody here knows nor cares about downstream.

PANAMA CANAL Central America

1910 The Panama Canal is one of the great engineering schemes of the 20th, or any other, century. This shipping route through Central America, from the Atlantic to the Pacific, cut 13,000km (8,000 miles) from a typical journey between the two oceans. But it was a huge endeavour. A French project to cut through the rainforest-covered mountains of Panama foundered in the 1880s with 22,000 workers dead, many from malaria and yellow fever. But US engineers took over and completed the 77km (48-mile) canal in 1914 – albeit at the cost of another 5,000 workers' lives. The most difficult task was creating the Culebra Cut – also known as the Gaillaird Cut after its chief engineer – through the mountainous central spine of the isthmus. This shot was taken at Empire, an engineering base on the Cut.

Troubled thoroughfare Here is the Culebra Cut today, after recent widening to allow two large ships to pass each other. Each year more than 14,000 ships carry approaching 300 million tonnes of cargo through here. The journey takes about nine hours. But some modern ships are too large for the canal. And the canal is also beset by droughts caused, it is believed, by deforestation of the isthmus. As its reservoirs empty, there is sometimes insufficient water in the dry season to keep the Cut full.

WADI AS SIRHAN Saudi Arabia

1986 Wadi as Sirhan is a shallow depression in the vast Saudi Arabian desert, close to the country's northern border with Jordan. The depression contains fertile alluvium, but the area is so arid that it has always been a barren wasteland occupied only by passing bedouin and their camels. But hidden beneath the desert lies one of the largest underground reservoirs of water on the planet. The water mostly fell there in wetter times, and much of it is more than 20,000 years old.

Greening the desert In the past two decades, the Saudi government began to bring this underground water to the surface to irrigate crops. Each green disc in this image is a field irrigated from a rotating sprinkler arm up to 300m (1,000ft) long. Most of the fields grow wheat. But growing wheat in the Saudi desert is hopelessly uneconomic, sustained only by heavy state subsidies and the free supply of water. Worse, it is very inefficient. Much of the water evaporates before reaching crop roots. Today, thanks to such projects, more than half the country's ancient and irreplaceable underground water reserves are gone. This is much to the anger of Saudi Arabia's Jordanian neighbours, because the rocks containing the water extend either side of the border, and Jordanians believe these Saudi fields are being irrigated with water sucked from beneath their soil.

MACHALA Ecuador

1991 As the oceans empty of fish the world is growing increasingly reliant on farming fish and crustaceans in ponds on land. Shrimp ponds are among the most lucrative forms of mariculture, and across the tropics huge areas of coastal wetlands have been converted to raising shrimps, often for sale to consumers in Europe and North America. Here in Ecuador, on the southern shores of the Gulf of Guayaquil near the town of Machala, the invasion was already evident in 1991. Farmers were chopping down mangroves and raising earth banks along the creeks to convert them to shrimp ponds. In this image, 143sq km (55sq miles) of coastal wetland is already occupied by the ponds.

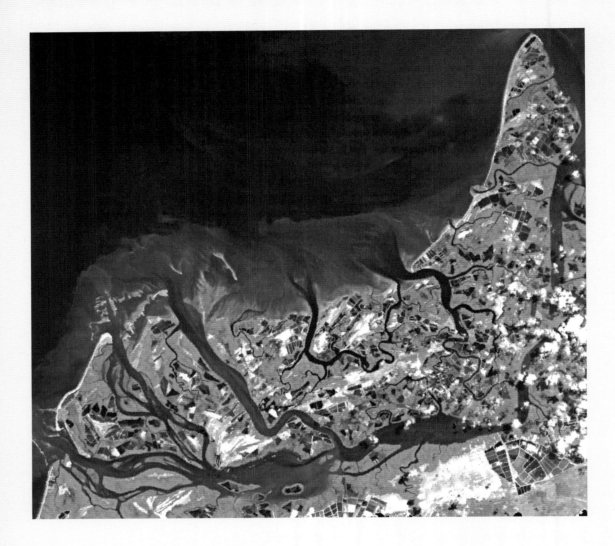

Coastline conversion But that was just the start. In the 15 years since the first picture was taken, a further 100sq km (40sq miles) have been converted. Fully 83 percent of the wetlands here have now gone. Ecuador has become a major exporter of farmed shrimps. They bring wealth to the country. But with the mangroves gone, the coastline is more vulnerable to storms and coastal flooding, ocean fisheries are disappearing because their nurseries among the roots of the mangroves have been lost, and the wider ecosystem is suffering. Ultimately, the ponds themselves may become barren.

ALMERIA Spain

1974 Almeria in southern Spain is arid. In the 1970s, spaghetti westerns were filmed here. Back then, farming was patchy and small-scale, largely for want of water. This image of the Campo de Dalias, south of the mountains of the Sierra Nevada, was taken in January and shows the land fairly green, but in summer it was yellow. Soon afterward, more intense irrigated farming began to develop, as farmers pumped water from underground and state authorities tapped rivers flowing through the mountains. And in 1986 Spain joined the European Union, opening up huge markets for agricultural produce that grew best in hot sun, such as tomatoes, oranges, and citrus fruit.

Plastic plague Today intensive farming has transformed the landscape of Campo de Dalias. The yellow and green has been replaced by the white of greenhouses glinting in the July sun. The plastic maximizes temperatures and boosts output through the year. Almost every patch of agricultural land is now irrigated and under plastic. The only areas spared are the mountains and the coastal strip, which is dominated by tourist developments. But the water is running out. Local underground reserves are drying up and plans to pipe water from Spain's wetter northern areas have foundered. Time for a revival of the spaghetti western?

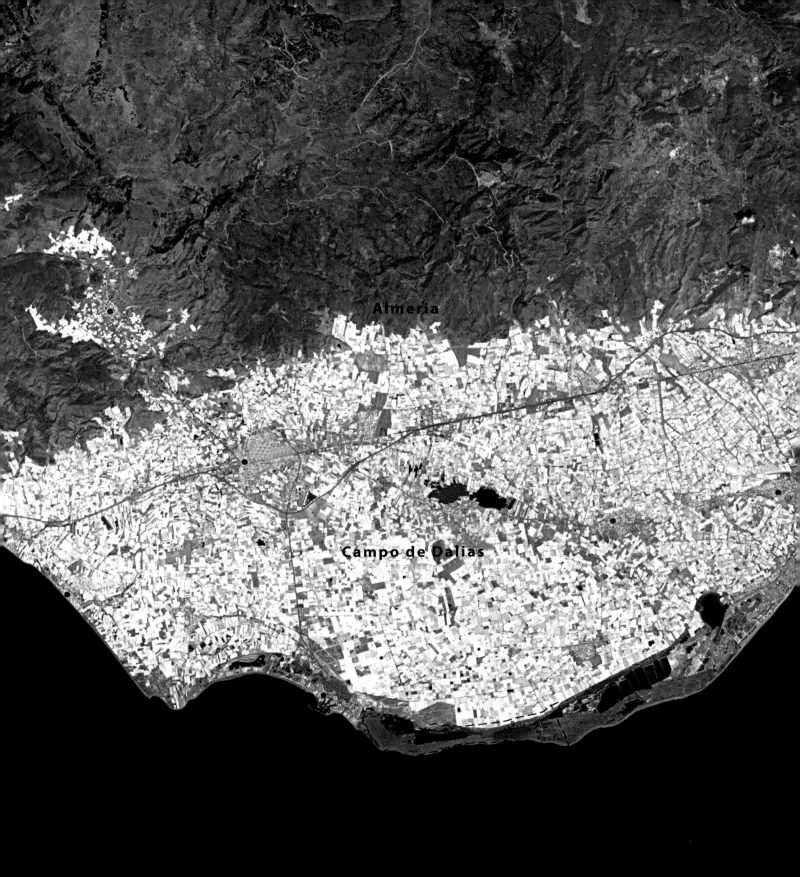

PARA Brazil

1990s Mangrove forests are among the most distinctive and ecologically important types of tropical rainforest. Made up of many species of salt-water tolerant trees and shrubs, they grow in the brackish inter-tidal waters of coastlines from southeast Asia, India, and West Africa to here in Para State in eastern Brazil, near the mouth of the great Amazon river. The trees sink their roots into the silt but draw much of their sustenance from the water itself. Mangroves are vital nurseries for fish stocks, nurturing fry among their roots. They also provide physical protection from coastal erosion and hurricanes by dissipating the immense energy from winds and waves. During the 2004 Indian Ocean tsunami, areas with mangroves suffered much less damage than those where they had been removed.

Mangrove clearing

Mangroves around the world have come under assault in recent years. Over half have been lost. The land and waters they occupy is often taken for commercial farming of fish and prawns. And their timber has many uses, from fishing traps and boats to making toilet paper for sale around the world. Here, mangroves on the banks of the Amazon in Para have been cut down to make fishing traps, leaving the land behind exposed to the full fury of nature.

RONDONIA Brazil

1975 Rondonia is one of the most remote regions of the Brazilian Amazon. High in the headwaters of the great rainforest river, it is 2,000km (1,200 miles) from the Atlantic Ocean. In 1975, as this satellite image shows, the forest canopy was virtually unbroken. The few small towns in the state, such as Rio Branco, were the homes of tribal communities and small bands of Brazil nut harvesters, many of them descendents of the region's rubber boom almost a century before. Crucially, there were no all-weather roads into the region from the great population centres of Brazil to the east.

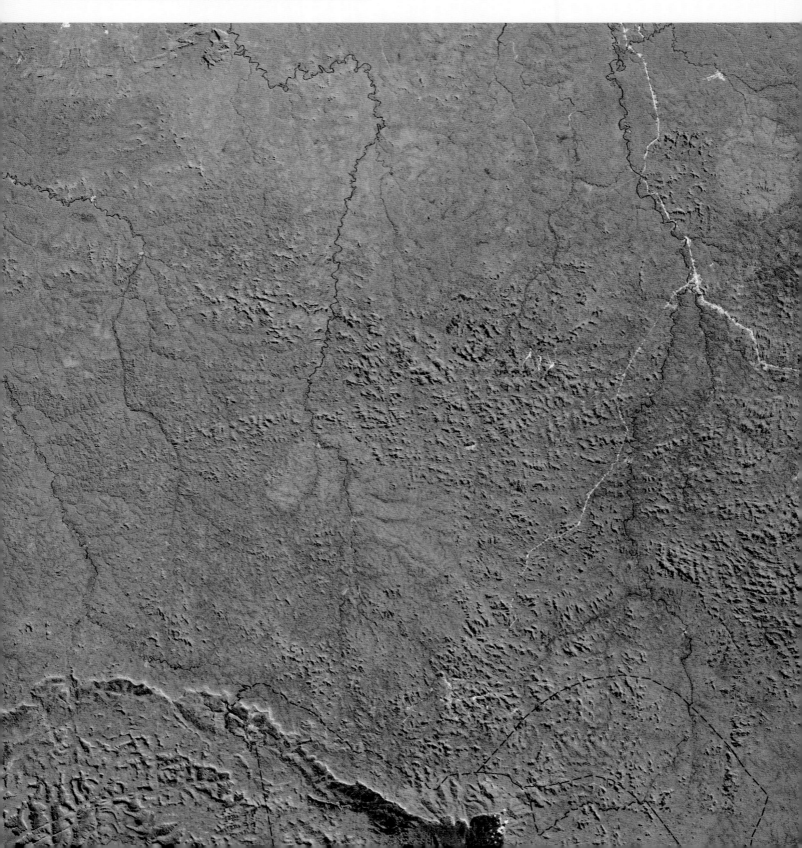

Deforestation

A quarter-century later, the pristine canopy of the forest has been lacerated by loggers. The paving of the BU-364 highway through the province brought loggers and thousands of migrant farmers from the east. Their activity is concentrated along the dense network of sideroads created by logging companies after the highway came through. The result, seen from the air, is a distinctive herring-bone pattern of deforestation. The only large surviving areas of intact forest are those protected as national parks or for indigenous tribes, including the areas in the bottom of the picture.

SUBTROPICAL FOREST South America

1973 Three decades ago this was a remote land of rainforests and rivers in the heart of Latin America, where Brazil, Argentina, and Paraguay meet. Ecologically, the most treasured area was the Iguazu National Park, a protected rainforest in the Argentinian sector in the bottom-right of the picture. Three small towns – one in each country –

hugged the river banks where the south-flowing River Parana met its west-flowing tributary, the Iguazu. As an ominous sign of things to come, an extensive herring-bone pattern of forest clearance for farms is evident on the new roads west of Ciudad del Este.

Three of a kind Today much of the area has been transformed. There are three giant hydroelectric dams in this picture. By far the biggest reservoir is the one behind the Itaipu Dam on the Parana, where it forms the border between Brazil and Paraguay. For 25 years, until the completion of China's Three Gorges Dam, Itaipu was the largest hydroelectric plant in the world, supplying power to Brazil's great industrial centres of Rio de Janeiro and São Paulo, as well as to Paraguay. At the top left is Paraguay's Acaray reservoir and, at the bottom right, Argentina's Urugua-i Dam. Meanwhile, the three towns have grown hugely. Ciudad del Este is now Paraguay's second-largest city with a population of a quarter of a million, mainly Asian migrants. And, while much of the Argentinian park survives, most of the Paraguayan forest has been converted to farming.

GIFFORD PINCHOT NATIONAL FOREST USA

1993 The Gifford Pinchot National Forest is one of the oldest protected areas of forestland in the US. It began life as the Mount Rainier Forest Reserve, created in 1897. It covers more than 5,300sq km (2,000sq miles) and includes within it Mount St Helens, the volcano that exploded in 1980. For many years it was not obvious why the great forests of the American West needed protection. They were vast and the local populations were small. Few besides miners and the engineers laying railroads disturbed the land. Then the logging companies arrived. And this image – a single photograph, though it appears almost to be a composite of two – reveals why forests need protecting. On the right is the Gifford Pinchot National Forest. On the left is a forest logging concession owned by the private company Weyerhaeuser, one of the largest in the American West. The company, entirely legally, has clear-cut the mountainsides right up to the edge of the National Forest.

BINGHAM CANYON USA

1964 The Bingham Canyon copper mine near Salt Lake City in Utah was opened in the early 1960s. Miners extracted the copper in layers, or benches, that they blasted in turn. The copper-containing rocks, known as ore, were loaded by giant shovels into electric train cars that ran on rails beside each bench. The ore was taken away for refining and smelting to create pure copper, one of the world's most widely used metals.

The big dig Bingham Canyon is today owned by the mining giant Rio Tinto. It is the world's largest man-made hole. With dozens of benches blasted into the Earth, it is now 3km (2 miles) across and more than 1km (0.6 mile) deep. Today's electric shovels can lift 100 tonnes of ore in a single scoop, and the dump trucks that have replaced the rail cars can carry 250 tonnes. Half a million tonnes of rock is cut daily from the mine, which has so far produced 17 million tonnes of copper, as well as by-products of gold and silver. The mine is likely to stay open until at least 2013.

ATACAMA DESERT Chile

1989 The Escondida Mine in Chile's Atacama desert is the world's most productive open-pit copper mine. It produces nearly 10 percent of the entire world's supply of the metal, as well as smaller amounts of gold and silver, and is owned by two of the world's largest mining companies, BHP Billiton and Rio Tinto. This aerial image was taken shortly before the first output from the new mine. It shows the mine itself, roads to the site (which is 3,000m/10,000ft above sea level), and, in the bottom left corner, the site of an intended reservoir for liquid waste from the metal-refining processes.

Digging deep The mine has been growing continuously. More than 120,000 tonnes of ore are removed from it every day. The ore is milled onsite to produce a copper concentrate that is mixed with water to make a slurry that can flow down a 170km (100-mile) underground pipe to the port of Coloso, where it undergoes further processing. The tailings reservoir now contains some three billion tonnes of waste sludge. It has recently been redeveloped to recycle water, the most precious resource at the desert mine. The mine's output is currently approaching a million tonnes of pure copper a year, which is used for everything from copper cables to plumbing to mobile phones.

CAPE YORK Australia

1973 Weipa on the Cape York peninsula in Queensland, Australia, is the site of one of the world's largest open-cast bauxite mines. It produces the raw material for aluminium, one of the most ubiquitous of metals, used in everything from drinks cans to jumbo jets. In 1973 mining activity here was just a decade old. Its scars are easily visible on the landscape round Albatross Bay, where the natural green vegetation has been cleared and the soils scraped away to reveal the bauxite just below the surface.

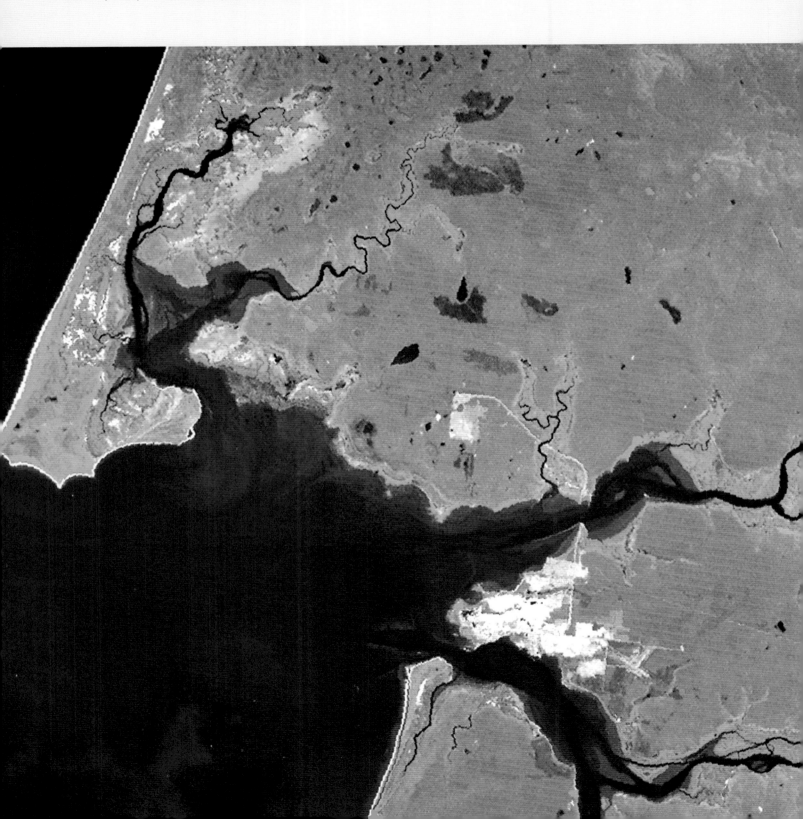

Mining ambition Thirty years later, the mining has spread its footprint much further. The original mines are partially restored. The land is returning slowly to its former green colour, though often several metres lower than before. But other mining areas have proliferated all around the bay. The bright blue area below Weipa town is settling ponds for waste. The mine today produces more than 15 million tonnes of ore annually, which is shipped around the Cape to Gladstone for refining and smelting. Mining has been completed on around 100sq km (40sq miles), and around 70sq km (25sq miles) has so far been revegetated. But this is just the start. The total area leased for mining covers some 2,600sq km (1,000sq miles).

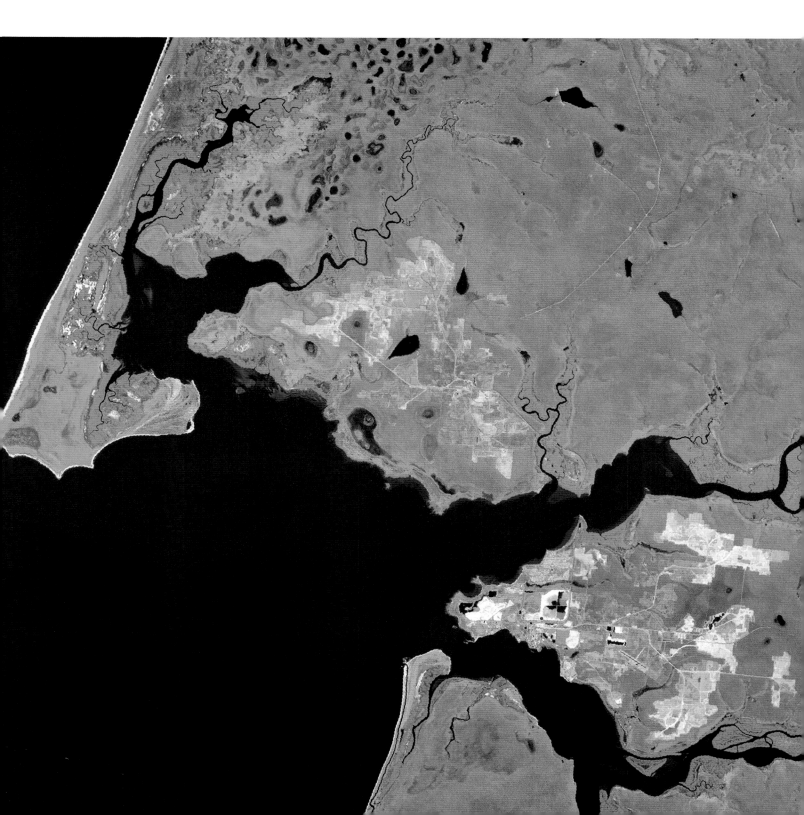

NAURU Pacific Ocean

1968 The tiny Pacific island of Nauru was covered in a thick layer of phosphate rock, the dried remains of thousands of years of bird droppings. Its only industrial activity for the past century was mining that phosphate to provide fertilizer for farms in Australia and elsewhere. So single-minded was the mining that by 1968, the year the island won independence, 17 of its 21sq km (8sq miles) of land looked like this: a moonscape of limestone pinnacles left behind after the removal of the island's only natural resource. The cranes sit on a small area not yet excavated.

Mined out The mining has all but ceased today. The phosphate is largely gone and vegetation is slowly returning to the moonscape. But after decades of providing fertilizer for the world's farms, the 13,000 inhabitants of Nauru have no soil themselves on which to farm. A £50-million payment from Australia in the 1990s to compensate for the destruction of their island made Nauruans among the richest in the world, per head of population. But the money has been frittered away in bad investments; Air Nauru's last Boeing was repossessed in 2005. Since 2001 Nauru has accepted Australian aid in return for being used as an offshore detention centre for asylum seekers. The phosphate island is now little more than an offshore prison hulk.

VANCOUVER Canada

1907 Early settlers to the Canadian Pacific coast called it the little mountain, a tree-covered granite outcrop in the midst of the flat plains that became Vancouver. But as the city grew, the 170m (550ft) forested hill became a vital resource. First it was logged for timber, and then at the start of the 20th century miners opened three quarries at its summit to provide crushed granite for construction of the first paved roads in the booming sawmill town. This picture was taken at the height of the quarrying era. But three years later the granite was exhausted and the quarries were closed.

The return of nature As Vancouver continued to grow, it encircled the pockmarked and deforested mountain. So the city fathers decided to create a park on its slopes, which they planted with all the native trees of Canada. Queen Elizabeth Park was opened in 1939 by King George VI and his consort Queen Elizabeth (the mother of the current Queen of England). At its centre they turned one of the pits into a sunken garden, seen here. The quarry gardens are now one of the top tourist attractions in Canada's third-largest city, drawing six million visitors a year.

ISAHAYA BAY Japan

1993 Japan's civil engineers like land reclamation projects almost as much as the Dutch. They say their mountainous but densely populated country needs as much extra land as it can get. Isahaya Bay on the island of Kyushu was a prime candidate. The bay is full of tidal flats that are barely flooded. In fact, it contains Japan's largest area of tidal flats.

As the 1993 image shows, farmland encroached right up to the shoreline, constrained by sea walls but seemingly eager to advance further. Small-scale land reclamation, launched in 1989, was already under way at this point.

Land reclamation
Major reclamation began in Isahaya Bay in the 1990s with the construction of a 7km- (4.5-mile-) long wall across the bay. It enclosed 3,000 hectares (7,400 acres) of tidal flats. By 2003 large areas of formerly flooded bay have been drained and turned into fields. The coastline has been transformed. Behind the wall, more tidal areas are slowly drying out ready for conversion. But not everyone is happy. Environmentalists say the old tidal flats were vital bird sanctuaries. And some farmers complain that they have lost valuable shallows where they once collected seaweed. There is now a campaign to tear down the new sea wall and restore the natural ecosystem.

PRIPYAT Ukraine

1960s Pripyat was never a pretty city. A metropolis of some 50,000 people, it was built by the government of the Ukraine when that country was part of the Soviet Union to house workers at a power station 3km (2 miles) away. That power station was Chernobyl, which suffered the world's worst nuclear accident in 1986. It took the Soviet authorities some time to acknowledge the scale of the disaster. But four days after the accident, with fallout raining down on Europe – its cloud stretched as far as the hills of north Wales – the population of Pripyat was finally evacuated.

A deserted city Pripyat has stayed evacuated ever since, along with a wide exclusion zone downwind of the plant. Radiation levels remain too high for people to return to this ghost city, which has become an unvisited museum of life in the old Soviet Union. It will probably be centuries before Pripyat is safe enough for people to return there without protective clothing. But, with humans virtually absent, nature has made a spectacular comeback. Trees grow in the radioactive soils, where children once played. Wild boar, deer, elk, beavers, and even wolves have invaded the city in their thousands. Unaware of the radioactivity that will doubtless shorten their lives, wild animals revel in the absence of humans. And behind it all the power station, with its core now in a concrete tomb, looms on the horizon.

ST PETER'S BASILICA, ROME Italy

1996 Smogs and other forms of air pollution have blackened and eaten away at the stonework and statues of most cities over the decades. The acidity of rain and fogs, which can contain moisture as caustic as battery acid, dissolves marble and limestone. And soot from smoke and diesel blackens everything. This image shows a travertine marble statue of an angel on the 400-year-old façade of St Peter's Basilica in the Vatican, damaged by centuries of exposure to the smoke of Rome, topped up by the diesel fumes of the tourist buses that visit the Vatican.

Papal restoration

After 30 months of restoration the detail of the statue is greatly improved, and hidden tints of red and green that were added to the marble in the 18th century are once again on full display. Pope John Paul II unveiled the restoration of the façade in late 1999. The work was sponsored by a major Italian oil company – whose products would have helped to cause the damage. It had taken $5 million and almost three years of scrubbing and cleaning to remove the stains from the façade, which is regarded as one of the masterpieces of early Baroque architecture.

KUALA LUMPUR Malaysia

2005 As cities of the developing world go, Kuala Lumpur, the capital of Malaysia, has relatively clean air. Most days its gleaming new high-rise buildings, such as the Petronas Twin Towers seen here, are visible with ease from across the city of 1.5 million people. Malaysia, which plans to be a fully industrialized country by 2020, has attempted to get its first-world clean-air legislation in place ahead of time.

Forest fires Legislation may be well meaning, but it can still be compromised. Kuala Lumpur is only 100km (60 miles) from the Indonesian island of Sumatra, which is clearing its rainforests faster than almost anywhere on earth. The aim is to make room for palm oil plantations, and the favoured method is by setting fire to the trees. During the dry season – and particularly during El Niños, when the rainforests dry out – these fires often rage out of control. They create huge palls of smoke that are no respecters of international borders. On this occasion, noxious fumes from the forest fires filled Kuala Lumpur for several days, and the city's inhabitants could only hold their breath until the wind changed.

Forces of Nature

Whatever mankind's impacts on our environment, the forces of nature are often of far greater ferocity. The shifting plates that make up the Earth's crust constantly cause earthquakes and unleash molten rock and ash from the planet's core through volcanoes. Sometimes those emissions can be enough to shroud the Earth in particles that cool the atmosphere for months or even years – as Mount Pinatubo in the Philippines did in 1991.

In the distant past, such events have been very bad news indeed for humans. When Mount Toba in Indonesia blew its top 73,000 years ago, the cold was so intense for so long that *Homo sapiens* almost died out. Most of the time, however, the danger to humans from these geological forces comes more immediately – from the destructive power of exploding mountains, shaking rock, flowing lava, and sometimes from the avalanches, tidal waves, and collapse of buildings that they unleash.

Volcanoes are a constant threat. In 1980 Mount St Helens in the USA lost 400m (1,300ft) from its summit during an eruption, creating the largest landslide in recorded history. Most of the Caribbean island of Montserrat has been out of bounds since its volcano blew in 1997, and the majority of its former inhabitants have never returned.

Earthquakes are, if anything, even less predictable. Usually they kill as shaking buildings collapse, as at Bam in Iran on Boxing Day 2003, where 26,000 died in the rubble. But in mountains, too, quakes can cause lethal landslides and avalanches. Most of the dead from a quake in the Pakistani foothills of the Himalayas in 2005 perished when rocks rained down on them. What happened to the former town of Yungay after a small earthquake in the Peruvian Andes was even more horrifying. And earthquakes on the sea bed can cause tidal waves that roar inland sweeping all before them. On another Boxing Day, in 2004, a quake beneath the Indian Ocean drowned more than a quarter-million people.

Floods, earthquakes, and volcanoes are killing more people each year. This is largely because our ever-more-crowded planet puts more and more people in harm's way. Most people live on coasts, vulnerable to floods, for instance. Some even live below sea level, trusting, like the people of New Orleans, to the protection of sea defences. Volcanic regions often have extremely fertile soils, enriched over centuries by lava flows, and they attract people. An added painful twist is that it is the poor in many cities that so often end up living on land vulnerable to floods and landslides. Even in known earthquake zones, few buildings are designed and built to resist shaking.

But there is a growing fear, too, that humans can sometimes trigger "natural" disasters. By deforesting hillsides and draining marshes, we are artificially increasing the intensity of river floods. Global warming may be increasing the risk of avalanches from melting ice and thawing permafrost in high mountains. Is that what happened on Mount Cook in New Zealand in 1991, and at Yungay in 1970? Climate change may also be increasing the intensity of hurricanes – a hot topic after Hurricane Katrina broke levees and unleashed unprecedented floods in New Orleans in 2005. When nature conspires with the hand of man to unleash such forces, we should take great care.

Japan reckons it is better prepared for earthquakes than any other country in the world. So the damage caused in 1995 by the quake in Hyogo province came as a shock. In places the shaking from the seismic waves triggered by the quake continued for three minutes. Neither buildings nor urban infrastructure could take the strain. The city's railway was tipped up as if it were part of a child's train set. When the shaking finally stopped some 6,000 people were dead, mostly here in the historic city of Kobe, where fire had broken out among the old wooden buildings.

SURTSEY ISLAND Atlantic Ocean

1963 Early one November morning in 1963, a crack opened in the Earth's crust on the bottom of the Atlantic Ocean, about 30km (20 miles) south of Iceland. Red-hot magma burst upwards into the ice-cold water, while water fizzed down into the crack. Fishermen began to notice black smoke and then a stench of sulphur. For months the eruptions continued on the sea bed. The sea boiled, ash was blown into the air, and lightning shot between clouds of volcanic ash, until eventually a new island began to solidify above the waves. The island of Surtsey, named after a mythical Norse fire-eating giant, was born.

New growth Slowly the eruptions ceased and the island began to cool. During Surtsey's first spring, scientists stepped ashore to find that wildlife had got there first – a lone fly and some seeds dropped by passing birds. The following year they found white flowers of sea rocket, which had sunk its roots into the ash. Lichens and mosses, insects, and even sea-bird colonies have followed. It is, some say, like the world being born anew. But for how long? When the eruptions stopped, Surtsey had a surface area of 2.7sq km (1sq mile), but the Atlantic swell is already eating away at its cliffs and beaches. The island has halved in size since 1967. Within a century or two it may disappear beneath the waves once more.

HEIMAEY ISLAND Atlantic Ocean

c1970 Heimaey Island is just off the south coast of Iceland. Its only settlement, and one of Iceland's biggest fishing ports, is Vestmannaeyjar, which sits in the shadow of a volcano called Eldfell. The volcano was once thought to be extinct, but as events at nearby Surtsey showed, the area south of Iceland is still volcanically very active.

Race against time Early one January morning in 1973, Eldfell began to spit showers of red-hot ash into the cold air and onto the outskirts of the town. By dawn, the island's population of 5,000 had fled to the mainland. Meanwhile, the eruption continued. Some 65 buildings disappeared beneath the black cinders, and another 300 burned to the ground. But, worse still, a slow tongue of lava continued to flow out of the volcano, threatening to block the harbour and destroy the town's economy. For four months volunteers worked in relays to spray cold seawater onto the hot lava to solidify it and slow the advance. Just in time, the volcano quietened. The harbour was saved and the fishing community could start to rebuild. And the millions of tonnes of ash they cleared from the town came in handy to make a second runway for the island's tiny airport.

MOUNT ST HELENS USA

May 1980 The symmetrical conical peak of Mount St Helens in the US state of Washington was a sacred place for generations of native Indians. But they feared it, too, because of its history of eruptions. At 2,700m (9,000ft), with a distinctive snow cap and surrounded by forest, it was one of the landmarks of the US's Pacific Northwest. By 1980 it had not erupted for more than a century, and the fears of the natives were forgotten as tourists, loggers, and nature-lovers encroached on the mountainside.

Massive eruption At 8.32am on 18 May 1980, after weeks of rumbles and emissions of ash, Mount St Helens erupted with huge force. The volcanologist David Johnston, who screamed into his radio "This is it!" as the mountain burst, was dead two seconds later, obliterated by an advancing wall of rock, ice, and trees. The eruption had been triggered by an earthquake beneath the mountain. The mountain's north face collapsed in an instant, creating the largest landslide in recorded history. It buried everything in its path to a depth of 200m (650ft) and more. The blast wave snapped fully grown trees 20km (12 miles) away; 57 people died. When the clouds of ash cleared, shocked onlookers saw that the snow-covered cone had become a flat-topped snowless stub with a peak 400m (1,300ft) lower than before. Two years later the US Congress created the National Volcanic Monument, a vast national park around the mountain, where nature is being left to recover as best it can.

MOUNT ST HELENS USA

1973 This infra-red satellite image shows Mount St Helens before the eruption: a white summit of snow and ice surrounded by muddy slopes and a wider mountain region of forest, which shows up green.

Fallen debris

Ten years later, this new image shows the rock, mud, ash, and trees from the explosion in shades of grey. They are smeared across mountains and valleys to the north of the volcano for more than 10km (6 miles). A series of lakes, shown black, has formed where rivers were blocked by debris, in places 170m (560ft) deep.

1983 Most volcanoes are brooding presences. There may be hundreds or even thousands of years between each explosive eruption. That is enough time for many of the obvious indicators of volcanic activity to disappear, though they are always liable to erupt again. Others return to life rather more frequently, though usually less violently. One such is Mount Kilauea on the southernmost island of Hawaii. It is one of the world's most permanently active volcanoes. This is Puu Oo, its easternmost vent, which began spewing hot lava down the mountainside in 1983, and has been erupting slowly ever since.

Hawaii **MOUNT KILAUEA**

Constant lava flow Here is Puu Oo again in 2001. So far, over more than two decades, the current eruption has extruded some 2cu km (0.5cu mile) of lava from beneath the earth and covered more than 100sq km (40sq miles) of Kilauea's southern slopes. In the process it has extended the island's shoreline and destroyed 200 houses, a visitors' centre, a 700-year-old Hawaiian temple, and 13km (8 miles) of roads. There are no signs of it stopping any time soon. This persistent non-explosive activity – and the fact that it creates shallow slopes rather than the steep slopes of a conventional volcano – makes Kilauea a favourite among volcanologists. It is known as the "drive up" volcano, because even its central crater can be reached by car.

MONTSERRAT Carribean

1993 Montserrat is one of the smallest of the leeward islands in the Caribbean, just 16km (10 miles) long. It is also one of the last surviving specks of the British Empire, ruled from London. In the early 1990s it had a population of 12,000 people, mostly living in Plymouth, the sleepy port capital on its southern shore. The island had a tourist industry and its land was fertile, in large part because of the volcanic soils provided by the island's largest geographical feature, the Soufrière volcano behind Plymouth. But in 1997 the volcano erupted for the first time in recorded history. A cloud of ash, pictured here in the first minutes of the eruption, rose into the air behind Plymouth…

Blanket of ash Within hours of the eruption, huge volumes of ash and boulders rained down on Plymouth. This picture was taken later the same day, as residents hurriedly left from the town jetty. By the time the eruption was over the whole southern half of the island was blanketed, and parts of the town were buried in ash more than 3m (10ft) deep. Plymouth's airport was destroyed. Today just 4,000 people remain on the island, all living in the north, which was previously largely uninhabited. The southern half is still out of bounds and regarded as permanently unsafe because of sporadic further eruptions.

MOUNT PINATUBO Philippines

1991 Satellite images can pick out many features of the atmosphere. This compilation of false-colour images taken from a satellite in late June 1991 shows the presence over the oceans of aerosols – tiny specks of soot, dust, and other particles suspended in the atmosphere. The brown areas show particles mostly in the lower atmosphere, while the whiter areas show a prevalence of higher particles. The satellite cannot see aerosols suspended over the land. This distribution is typical for the time of year. Aerosols are concentrated close to the equator because most of the pollution comes from farmers burning fields and forests in Africa and Asia, and from dust storms from the Sahara.

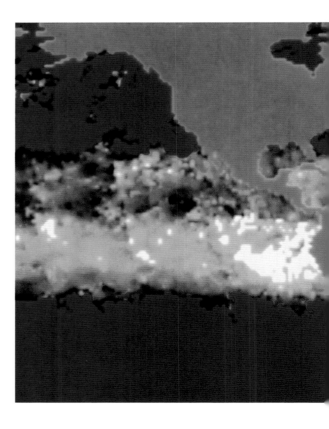

Volcanic plume Six weeks later, the narrow bands of aerosols close to the equator have been replaced by a thick and continuous plume stretching right around the Earth and from the north coast of Africa almost to its southern tip. The plume comes from Mount Pinatubo, a volcano in the Philippines that exploded on 16 June with such force that 150m (500ft) disappeared from the top of the mountain. Some five billion cubic metres (180 billion cubic feet) of ash was blasted high into the upper atmosphere – more than from any eruption since Krakatoa in 1883. The ash remained aloft for several years, scattering so much sunlight that it cooled the earth by about 1°C, the only serious interruption to global warming in the past three decades.

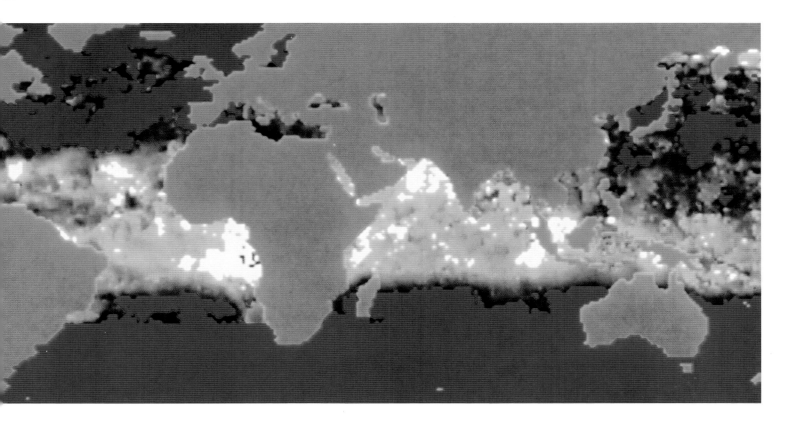

MOUNT COOK New Zealand

1991 Mount Cook is the tallest peak in New Zealand. Its sharp summit is also known by its traditional name of Aoraki, or cloud-piercer. The mountain is wrapped on all sides by two large glaciers that make it an attractive challenge for mountaineers, but also a hazardous one. The steep slopes are made unstable by the ice hanging from them. It could fall off at any time, perhaps taking rock with it.

Mountain-top avalanche On 14 December 1991

that is what happened. That night mountaineers waiting in huts for a morning ascent heard great rumbles from the mountain and saw sparks as rock clashed against rock in the dark. Something had given way on the summit and by daylight 14 million cubic metres (500 million cubic feet) of rock, ice, and snow had tumbled down from its east face. Some of the rock ended up 7km (4 miles) away on the Tasman Glacier below the mountain. The avalanche dramatically changed the shape of the peak and reduced its height by about 10m (33ft), to 3,754m (12,316ft). Further avalanches are expected, and many believe that global warming is likely to destabilize many more ice-covered mountain summits around the world.

BAM Iran

c2002 Bam is one of the great historic cities of Iran. It was founded more than 2,000 years ago by the Parthian Empire. Close to the border with Pakistan and due north of the Persian Gulf ports, it was on several major trading routes between the Middle East and Asia. The city's old town, seen here with the citadel at its heart, also housed a major Zoroastrian temple visited by thousands of pilgrims every year. The buildings were mostly at least 500 years old and made of adobe (sun-dried clay bricks), straw, and the trunks of palm trees, making Bam among the world's most distinctive cities.

Devastating quake On Boxing Day 2003, Bam's old town was flattened by an earthquake. Its mud buildings were shaken to pieces, its citadel largely wrecked, and some 26,000 people across the city died. Few survived because the brick buildings disintegrated entirely, creating a mountain of rubble. There were relatively few half-collapsed structures containing air pockets within which people could survive, as often happens in modern cities. An ancient city that had often been called the largest adobe structure on earth had been destroyed.

ANCHORAGE, ALASKA USA

1959 Anchorage began as a base for construction of Alaska's first railway. It was built where the railway met the sea, with the first streets laid out in 1914. For a long time it had no agreed name. Eventually, the US Postal Service gave up on locals and declared it to be Anchorage. Urban expansion began seriously in the 1940s with the construction of an air force base nearby. The town gained further importance in 1959, the year this picture was taken on Fourth Avenue, when Alaska was incorporated as a full-scale state within the Union.

After the quake On 27 March 1964, Good Friday, the five-year-old state
of Alaska was hit by one of the biggest earthquakes of the 20th century. It registered a
magnitude of 9.2. The change to the landscape from this huge shift in the Earth's plates
was immense. This photo shows Fourth Avenue two days after the quake: the road has
collapsed so completely that cars have ended up some 6m (20ft) below the former street
level. Yet, amazingly, while one side of the avenue was wrecked, the buildings on the
other side are largely intact. Altogether 131 people died. High-rise construction is still
banned here because of the quake threat.

YUNGAY Peru

1968 Yungay was a town in the Peruvian Andes with about 25,000 inhabitants, nestled in a lush green valley with spectacular views of the glaciers on Mount Huascaran. About 15km (9 miles) away and 3km (2 miles) higher than Yungay, the mountain is one of the grandest peaks of the Andes. On Sunday 31 May 1970, most of the population of Yungay were out buying produce at the Sunday market after church. There was a circus in town entertaining the children. Then there was a small rumble in the distance.

Killer flow An earthquake had occurred somewhere in the mountains above
the town. It destabilized the ice cap on the summit of Mount Huascaran and a wall
of ice fell down the slope. It probably did not attract much attention down in Yungay.
Not immediately, anyhow. But the wall of ice landed in a lake at the foot of one of the
mountain's glaciers. It set off a tidal wave across the lake that burst its banks. Millions
of tonnes of mud, ice, and water set off down the valley at a speed estimated at more
than 200km/h (125mph) – enough to catapult the wave of debris over a hill, into the
air, and down onto Yungay. Virtually the entire town disappeared forever beneath
the debris, taking around 23,000 people with it.

MAKHRI Pakistan

2002 The mountain villages of Pakistani Kashmir, in the foothills of the Himalayas, are among the most remote areas on earth. The villages huddle in steep valleys and it can take many days to reach them from the outside. The people are extremely poor. They have few roads, schools, or health clinics and only sporadic contact with any form of government. Here, photographed by satellite in September 2002, is the village of Makhri. It sits on a rare area of flat, fertile land beside the River Neelum, a tributary of the River Indus, and faces a tall mountain ridge on the opposite bank.

Crippling landslide In October 2005, an earthquake shook these mountains. Besides levelling buildings, it caused hundreds of landslides on the steep mountainsides that engulfed entire villages and blocked the roads that relief workers attempted to take over the following days. One of the worst slips saw the collapse of the western face of the mountain opposite Makhri. The landslide, visible here as mounds of grey rock and mud, blocked the large meander in the River Neelum, forcing it to take a new course. The formerly blue river is now chocolate brown with the accumulated debris of this and other avalanches. Some 200 people died here.

SCARBOROUGH England

1993 Holbeck Hall in Scarborough, England, had a venerable past. Built in 1880, the four-star cliff-top hotel, with its wide green lawns and grand views over Scarborough Bay, was a popular holiday destination for visitors to the Yorkshire seaside resort. But its misfortune was to stand on top of the town's south cliff, one of the fastest-eroding cliffs on the east coast of England. The clays and soft sandstone are no match for the waves pounding the bottom of the cliff. And coastal geologists say sea walls built elsewhere along the coast have starved the sea of sediment and increased its power to erode anywhere that has been left undefended. On 3 June 1993 the buildings were still 65m (215ft) from the cliff edge, but over two days and nights the cliff began to give way until Holbeck Hall stood on the very edge.

Coastal retreat One more night did for the hotel. The occupants had enough warning to leave, as the lawns buckled and the building creaked and cracked. Finally the ground beneath the seaward wing gave way and large parts of the Hall tumbled down the slope into the sea. From Scarborough south to the Humber estuary, the coast is in retreat. Geologists say 10 villages along a 60km (40-mile) stretch of coastline may have to be abandoned as the North Sea eats up the coastline at a rate that averages 10m (33ft) a year, but which some nights can be many times greater.

TWELVE APOSTLES Australia

2005   These rugged limestone stacks, known as the Twelve Apostles, stand in the sea within the Port Campbell National Park, 220km (135 miles) southwest of Melbourne in Australia. They are among Victoria state's top tourist attractions, easily visible from the spectacular Great Ocean Road. They stand as high as the cliff tops close by, an apparent symbol of the enduring ability of rocks to withstand the constant battering of ocean waves. But in fact their name has long been out of date. There were only nine stacks remaining when this photograph was taken…

Instant erosion …And less than one minute later, when this second photograph was taken, there were only eight stacks. The 50m- (165ft-) high stack in the foreground on the left had collapsed into the sea before stunned onlookers – a testament this time to the power of the ocean. Will the apostles eventually all disappear? It might seem so. But the land behind them is also being eroded, and as the sea gnaws away at fractures and other weak spots in the cliffs, leaving behind the hardest rock, new stacks could be created at any time. In a few decades, say geologists, there could be 12 apostles again.

ACEH Sumatra

2003 People have lived along the coast of Sumatra for thousands of years. They have long since cleared most of the forests and mangroves to make way for fields and rice paddy. But Aceh Besar, the northernmost district of Aceh, the northernmost province on the island, nonetheless remains largely rural. This patch of lowland, photographed in early 2003, is a lush green backwater in every sense. The rice paddy shows in light green. The water on the bottom left of the picture is a small freshwater wetland about 10km (6 miles) inland from the shores of the Indian Ocean.

Tidal wave On Boxing Day 2004, a tsunami struck the Indian Ocean. An underwater earthquake, the largest anywhere in the world for 40 years, caused devastation right around the Indian Ocean, but nowhere suffered more than Aceh, where an estimated 130,000 of the tsunami's quarter-million victims died. Within minutes the landscape was dramatically transformed. Here, the sea swept across the coastal town of Iho-nga and inland, engulfing the small wetland and consuming all the low-lying paddy and many settlements beyond. A few homes and roads were left marooned on higher ground.

ACEH Sumatra

2003 The lush landscape of Aceh is a rich tapestry of flooded rice paddy and prawn farms (the dark fields in the picture), woodland, and green fields. The shoreline – once fringed by mangrove swamps but now exposed to the waves – lies to the north. And a small river flows through the top right-hand section of the picture, where a few natural coastal wetlands survive. This is a densely populated, low-lying area facing directly into the Indian Ocean – and therein lies its vulnerability.

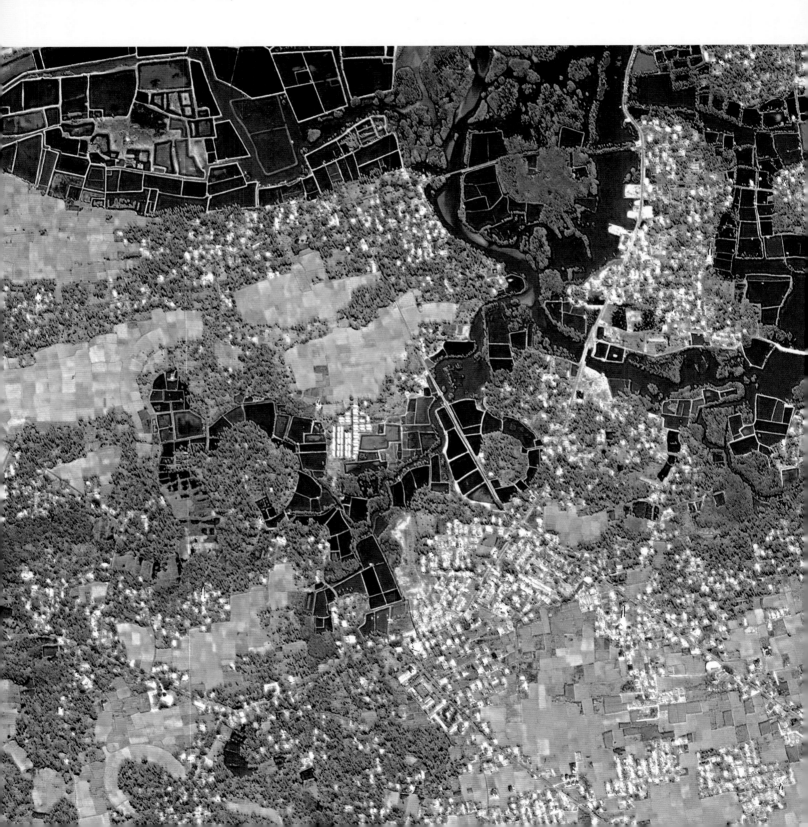

Catastrophic flooding The landscape has been transformed. This picture of the battered coastline was taken on 29 December 2004, three days after the tsunami struck. The shoreline paddy and prawn farms and many low-lying areas remain engulfed by the sea, which will leave behind a layer of salt that will prevent crops from growing. Whole neighbourhoods have yet to emerge from beneath the tidal wave that washed through the streets 72 hours before. A vast new river stretches from the bottom left to the top right of the picture. Only a few raised causeways stand out above the floods.

NEW ORLEANS USA

March 2005 New Orleans used to look like any other great city in the southern USA: all high rise and elevated freeways, the distinctive white roof of its Superdome glinting in the bright sun. But in the distance is a clue to its nemesis. The city is built behind levees on the delta of the Mississippi River, and the river and the ocean are higher than the city. Throughout its 300 years of growth, since its founding by French settlers, levees have been New Orleans' only defence against inundation.

Storm force And a feeble defence they proved in August 2005, when Hurricane Katrina roared in from the Caribbean and sent a tidal surge into the delta. It overwhelmed the levees and the result was this: highways turned to rivers, bridges marooned like islands, and whole neighbourhoods roof-high with the ocean. The cost of repairing the damage is now put at over $100 billion. And the human cost was equally great. In the middle distance, the picture shows the damaged Superdome, where 25,000 of the city's poorest inhabitants sheltered for days, during which nobody brought food or collected the dead. In all, some 1,500 people died in the disaster.

NEW ORLEANS USA

2005 The suburbs of New Orleans suffered as much as the city centre when Hurricane Katrina swept through southeastern Louisiana. This is the damage fully a week after the levees on the Industrial Canal, a major artery between the Mississippi and one of the delta's main lakes, were breached in three places. The neighbourhood is near the boundary between the city and St Bernard parish, just east of the downtown area. Only the roofs of houses are visible. Anyone who did not manage to get out or scramble onto a roof is dead.

Flood aftermath This is the same area, 12 days later. The waters are retreating. Some of the roads are dry, though others are still flooded. What remains of the houses is emerging, waterlogged and full of mud. A few people return to collect any belongings that survived. But this is a rare example of almost an entire major city – more than 80 percent by most counts – being inundated.

BETSIBOKA RIVER Madagascar

2003 The Betsiboka is the largest river in Madagascar. It drains a swathe of the island's mountainous north, an area once covered by rainforests containing wildlife unique to the island, such as its famous lemurs. This picture, taken from the international space station in September 2003, shows the river in the dry season. It is draining through a large delta into the ocean, which can be seen on the top-left. Upstream, largely out of shot, the rainforests have mostly been chopped down.

Cyclonic effect The second picture shows the same scene six months later, a fortnight after tropical Cyclone Gafilo swept across Madagascar bringing heavy rains. The denuded hillsides are now hugely vulnerable to erosion after rain. And the evidence is here. Rapid runoff from the hillsides has flooded the river and brought down millions of tonnes of soil. The soil has clogged the delta, stained the river bright orange, and sent a vast plume of sediment into the estuary. It is as if the island is bleeding into the ocean. The constant arrival of silt down the river is clogging up the estuary so that ocean-going ships that could once travel upstream must now berth on the coast.

MISSOURI RIVER USA

Early 1993 The Missouri river in the US winds across a vast, flat floodplain before draining into the Mississippi to form the second-largest river drainage catchment on Earth. The floodplain was once dotted with lakes and marshes that took the excess flow after heavy rains. But none are left in this picture. The floodplain has been drained and turned into rich arable farmland protected by levees. In the past century, some $7 billion has been spent raising levees along the Mississippi and Missouri to keep the water off the land. And, as this picture shows, farmers were confident enough of the protection they provided to plant their crops almost up to the river's edge.

Burst banks Levees can prevent small floods, but when they burst the results are
catastrophic. So it was in 1993, when some 1,000km (600 miles) of levees were breached.
Here, just weeks after the first picture was taken, the Missouri has flooded across farmland
for several kilometres either side of its normal channel. Some called it a natural disaster.
Others pointed out that the river was only flooding its floodplain: the disaster was
man-made. After the 1993 floods, there were calls to give more space back to the
river. But little had been done by the time of Hurricane Katrina in 2005, when again
the river reclaimed its floodplain, with disastrous consequences for New Orleans.

SOUTH DAKOTA USA

1935 The "dust bowl" took a terrible toll across the American Midwest in the 1930s. When drought hit the dry plains, crops failed, fields turned to deserts, the skies filled with clouds of dust, and millions of destitute farmers headed west to become labourers in factories and plantations in California. John Steinbeck's epic tale *The Grapes of Wrath* described the disaster best. Here, at the height of the drought, sand dunes up to 2m (7ft) deep have formed across formerly ploughed fields and encroach on a farmstead in Beadle County, South Dakota.

Soil conservation Many believed that the "desertification" of the Midwest would be permanent. But the federal government set to work with a soil conservation programme, part of the New Deal for America's poor. It achieved extraordinary results. Here is the same farmstead two years later. An emergency cover crop of Sudan grass has captured moisture and stopped the soil blowing away. Later, boreholes were sunk to abstract copious amounts of underground water from beneath the High Plains. Now farmers can irrigate their crops during dry spells. There's just one problem: the aquifers are emptying, so the region may again be living on borrowed time.

ATACAMA DESERT Chile

Early 1990s The Atacama Desert in northern Chile is the driest place on Earth. Average rainfall is just 1mm a year, but many places go for many years – in one recorded case for 40 years – without a drop of rain. The desert stretches for 1,000km (600 miles) parallel to the Pacific coast, in a rain-shadow between coastal mountains and the high Andes. The Atacama's salt plains and lava flows are about the most inhospitable place for life on Earth. Recent research suggests the soils are so devoid of microbes that they are more like those on Mars than any on Earth.

Desert blooms Just occasionally it does rain in parts of the Atacama, and the results can be extraordinary. Even if the rains come only once a decade, seeds that have remained dormant in the ground for years can germinate within hours and burst forth in amazing carpets of colour. Species often flower in sequence, so that sometimes a valley will be purple one day, yellow the next, and blue after that. Usually these desert blooms appear with the occasional heavy rains caused by El Niño, a periodic reversal of normal climatic patterns in the Pacific.

War and Conflict

Arguably, wars have disfigured the modern world even more than the destruction of the natural environment. The 20th century may one day be known more for the extraordinary carnage of its wars than for anything else. This chapter looks at the physical impact of the industrial-scale annihilation of people – but also at subsequent efforts at reconstruction and commemoration.

Look at the remains (or rather the absence of remains) of a Belgian village called Passchendaele, where three quarters of a million soldiers died over five months during World War I in an ultimately meaningless contest for the village. Or try to imagine the human tragedy behind the simple street scene in St Petersburg, where half a million died of starvation and disease during a 900-day siege during World War II.

Wars mostly kill people, but sometimes the environment is directly attacked for strategic advantage. In this chapter we show two examples of the latter. One is the extensive damage done to the rainforests of southeast Asia by the relentless US bombardment with the herbicide Agent Orange. Its purpose was to expose the jungle supply routes of the Vietcong during the Vietnam War. Another is the deliberate draining of the Mesopotamian marshes – the largest and most fecund wetland in the Middle East – by Saddam Hussein during the 1990s. His purpose was to force out the rebellious Marsh Arabs, who had occupied the marshes for thousands of years.

Nuclear weapons are indiscriminate destroyers of the environment and humanity. The obelisk in the New Mexico desert doesn't look much, but it was here that the first atomic bomb was tested in July 1945. The success of that test not only stunned the creators of the bomb, it also ensured the destruction of Hiroshima and Nagasaki in Japan – and the loss of 200,000 civilian lives – less than a month later.

But maybe this chapter can offer some hope. For it shows the power of the human spirit to heal wounds. Looking at the pictures of Caen and Dresden today, it is hard to imagine that either city was horrifically firebombed little more than half a century ago. Even Hiroshima has recovered, although its oldest citizens continue to die from diseases caused by the radiation released the day that "Little Boy" was detonated over the city.

The barricade at the Brandenburg Gate in Berlin was a symbol of the Cold War, and its dismantling has become a symbol not just of the reunification of Berlin and of Germany, but of the reunification of Europe and the end of the most dangerous stand-off in human history. Not that Europe has been entirely at peace since. It has been scarred by a decade of war in the Balkans following the collapse of Yugoslavia – a complex series of conflicts symbolized here by the death and rebirth of the Mostar Bridge in Bosnia.

And the Middle East continues to fester. Witness the bombings of Beirut and the construction of Israel's security barrier, which keeps out would-be terrorists but at the expense of any hope of peace between Israel and its neighbours. Many believe that, while that conflict persists, the fuel exists for a wider conflagration between the Islamic world and the West. And in that dispute nobody is safe – as workers at the Twin Towers in New York discovered one bright sunny morning in September 2001.

Many thought the atom bombs that obliterated Hiroshima and Nagasaki were the ultimate weapons of war. But on 1 November 1952 the US detonated an even more fearful device: the hydrogen bomb. This is the mushroom cloud of the 10.4-megaton explosion of Ivy Mike, which took place on Elugelab island in the Pacific Ocean. It created a crater 1,900m (6,240ft) in diameter and 50m (160ft) deep, and entirely vaporized the island.

PASSCHENDAELE Belgium

1917 Few place names chill the bones of historians more than Passchendaele. It has become a synonym for the horrors of war. In 1917, at the height of World War I, British, Canadian, and Australian soldiers wanted to cut through German lines in Belgium to capture a German submarine base on the coast. They chose to do it at Passchendaele, a small Belgian village in a marsh near Ypres.

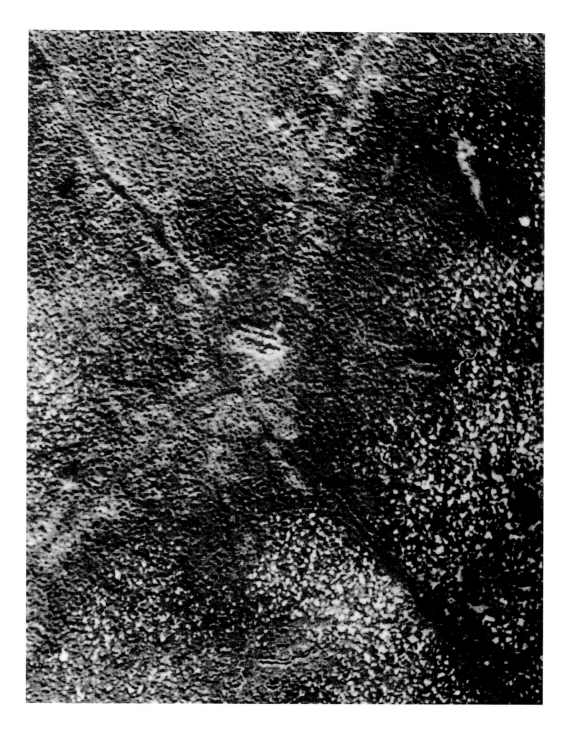

Obliteration The battle lasted five months. Tanks, mines, and artillery were deployed on a scale never seen before. So was mustard gas. The constant bombardment turned the marsh into a huge quagmire. Many soldiers drowned in the mud after falling from duckboards. The village was captured, and this is what was left. Nothing. The destruction was as total as if an atomic bomb had struck the village. Aerial photographs such as this showed a million shell holes within just over 2sq km (0.8sq mile). That is one for every 2sq m (22sq ft) of sodden ground. The Germans lost 270,000 men, and the British and their allies not far short of half a million. Some 90,000 bodies were never identified, and 42,000 never recovered. The war poet Siegfried Sassoon wrote: "I died in Hell; they called it Passchendaele."

THE SOMME France

1916 It was the first day of the Battle of the Somme, one of the most bloody conflicts of World War I. At 7.30am on 1 July the soldiers of the Newfoundland regiment, who had just arrived from Canada, left their muddy trenches dug into French soil at Beaumont-Hamel and advanced on German lines. Officers told the young recruits that the Germans had been decimated by five days of artillery bombardment. Instead, their opponents had survived the onslaught relatively unscathed, and cut them down with machine-gun fire. Of the 800 Canadians who went "over the top" that day, 310 died and only 110 returned unhurt – just one skirmish on a day when 20,000 died along the 45km (30-mile) front. Altogether, the Battle of the Somme lasted five months and when a halt was finally called, more than a million had died.

Remembrance Today Beaumont-Hamel is once again a quiet hamlet, 9km (6 miles) north of the town of Albert in northern France. But on its outskirts, the trenches and artillery-blasted fields are a reminder of what happened there. And the 30 hectares (75 acres) where the battle was fiercest is one of the most visited of all the memorial grounds for the dead of the Somme. The Beaumont-Hamel Newfoundland Memorial Park, opened in 1925, contains three cemeteries with 630 graves. And beneath the grass lie the unmarked remains of the 90 men who fell that day but whose bodies were never recovered.

LONDON England

1940 Cannon Street, in the heart of the City of London, England, is known today mainly as one of London's railway termini. But on 29 December 1940 it was firebombed by German aircraft at the height of their Blitz on the capital of the British Empire. After that night, and a further bombardment four months later, the street was reduced to rubble, as this view here from the top of St Paul's Cathedral, which miraculously survived, shows. In the distance, another survivor was Tower Bridge, over the River Thames.

In business Here is the same view today, with Cannon Street rebuilt and Tower Bridge still visible in the distance. Sixty years on the British Empire may have gone but the City of London stands beside New York as one of the two primary financial capitals of the world. Unusually for a major central business district, most of the new buildings are on a similar scale to those they replaced. What is different is their colour. The wartime buildings were blackened by decades of coal smoke that made London one of the smog capitals of the world. Vehicle fumes still fill the streets but the black coal smuts are gone and the new buildings stay the colour of their stone.

CAEN, NORMANDY France

1944 The Battle of Caen, between advancing Canadian forces and Germans in slow retreat, was one of the centrepieces of the final phase of World War II. It began days after the Canadians came ashore during the D-day landings on the Normandy beaches, and lasted for two months. The city's medieval heart was burned out when a rain of bombs on 6 June 1944 set the city burning for 11 days. As the Germans continued their resistance, most of the citizens took refuge in a few buildings, including the churches, that stood like beacons amid the rubble.

Reconstruction The heart of Caen has been almost completely rebuilt since the War. The churches and the ancient castle ramparts are almost all that remains of the old city. But some of the original streets live on as modern shopping precincts, cobbled and pedestrianized. The bunting in this picture celebrates the city's survival, and displays the flags of the liberating forces. Old soldiers and tourists alike use the hotels as a base to visit the famous landing beaches along the shore to the north of the town.

HOHENZOLLERN BRIDGE, COLOGNE Germany

1942 Intense Allied bombing in 1942 destroyed much of Cologne, Germany's fourth-largest city, causing severe damage to the famous Hohenzollern Bridge across the River Rhine. Behind it, the twin-spired cathedral, one of the largest Gothic buildings ever constructed, was almost the only building in the city's heart to survive. The bridge was swiftly repaired. But in March 1945, with the Allied forces sweeping east, the German forces abandoned Cologne and blew up the bridge as they went. In a last, desperate effort to halt the Allied advance, they destroyed most of the other bridges across the Rhine between Bonn and the Netherlands as well. But, in the process, they condemned to capture some 50,000 German troops stuck on the west side of the river.

The new bridge The bridge has been rebuilt once more. While earlier versions carried road as well as rail traffic, the post-war version is reserved for trains and pedestrians. But it remains a central link between the city's east and west sides, and a main route from Cologne to its hinterland on the western side of the river.

FRAUENKIRCHE, DRESDEN Germany

1952 The Allied firebombing of Dresden back in 1945 was some of the most intense in World War II. Much of the southeast German city was flattened – including the totemic Dresdner Frauenkirche, a baroque 18th-century Lutheran church in the heart of the city. The church survived 48 hours of bombing, but finally succumbed to temperatures of up to 1,000°C (1,800°F) created by the fires that raged through the city. Its architectural crowning glory, a 12,000-tonne sandstone dome held up by eight pillars, gave way when the pillars glowed bright red and exploded.

Reconciliation In communist East Germany the ruins were left untouched as a war memorial until the early 1980s, when the site became the focus for a civil rights movement protesting against the government. It became one of the sparks for the reunification of Germany. After reunification, an international movement developed to reconstruct the church, as an act of reconciliation between once-warring nations. Reconstruction began in 1993 using the original plans and some 8,000 of the original, charred stones, which are clearly visible here. The exterior was completed in 2004, and the church reconsecrated the following year. Once a month an Anglican Eucharist is held.

LENINGRAD Russia

1942 The Siege of Leningrad was one of the worst horrors on the Eastern Front during World War II. This image says much about how the extraordinary can become normal in times of war. A woman pulls a shrouded corpse on a sledge down snow-covered Nevski Prospekt, the city's main street – the event goes barely acknowledged by passers-by. The siege lasted for 900 days. The first winter, when this picture was taken, was one of the coldest of the century. An estimated quarter of a million citizens died of cold and starvation. And, altogether, half a million may have died before the Germans gave up and went home in early 1944.

Metamorphosis Much has changed. The Germans have long gone, but so, too, has Communism. The city has reverted to its pre-Soviet name of St Petersburg and this is the new-look Nevski Prospekt. The huddled masses are replaced by Japanese cars and Western-style advertising on what the city now hails as "Russia's most famous street", the home of Tchaikovsky and Nijinsky. What has stayed the same is the street's buildings, most of which date back long before Lenin, to the days of the czars. The winters are mostly warmer now, too.

TARAWA Pacific Ocean

1943 Tarawa is a clutch of 24 tiny coral islands huddled around a sleepy lagoon right on the equator in the central Pacific – an atoll that became a battleground in World War II. Here is the scene in November 1943 as US forces fought to retake the islands from some 5,000 Japanese, who were heavily dug in. The marines secured the atoll, but only after three days of fierce fighting that left the beaches covered in dozens of bodies and huge amounts of military hardware.

Battle scars Tarawa was, until 1979, the capital of the British-owned Gilbert & Ellice Islands, although London rules counted for little in wartime. Now it is the capital of the independent Republic of Kiribati, whose main source of revenue is tourists and revenues from selling rights to fish in the vast swathe of the Pacific that it controls. But, however much its politics have changed, many of the remnants of its brief moment of infamy still litter the beaches. Far from detracting from its value as a tropical paradise, the remains of the Battle of Tarawa and the Japanese defensive installations are today one of Tarawa's principal tourist attractions.

TRINITY SITE, NEW MEXICO USA

1945 Before the USA dropped its atomic bombs on Hiroshima and Nagasaki in Japan, at the end of World War II, it tested a prototype device in the US desert. Here is the site of the world's first atomic bomb test. Called the Trinity site, it is 300km (185 miles) south of Los Alamos in New Mexico. The tracks lead to where the bomb, hanging from a 50m (165ft) tower, was detonated at dawn on 16 July 1945. It exploded with the force of 19,000 tonnes of TNT, instantly vaporizing the tower and creating a crater, which appears small in this picture but is 300m (985ft) across and 3m (10ft) deep. The resulting mushroom cloud reached 12km (7.5 miles) into the stratosphere. Nothing survived within 1km (0.6 mile) of the blast. Observers 9km (5.5 miles) away were knocked to the ground by the blast wave. The sand around the crater was melted to a green radioactive glass that shows dark in the area surrounding the crater. They called this new substance Trinitite.

Monument In 1952 military engineers bulldozed the site of the explosion and buried the Trinitite, although collectors removed some samples first. Ground Zero is marked today by an obelisk that visitors are allowed to see twice a year. Radiation levels remain ten times the normal level. This is the site where the atomic age began and the world saw the first mushroom cloud.

NAGASAKI Japan

1945 Nagasaki was a pleasant backwater – a district capital with a population of around 200,000 on the Japanese island of Kyushu. It was not unknown to the outside world, having been the setting for Puccini's opera *Madame Butterfly*. It also had a deep harbour and became a major Japanese naval base in World War II. Shown here is the suburb of Urakami, beside the Urakami river in the north of the town. It was the heart of Nagasaki's Catholic community, and home to a medical college and several schools as well as a munitions factory.

After the bomb

An atomic bomb, known to the US aircraft crew carrying it as "Fat Man", was detonated over Nagasaki at 11.02am on 9 August 1945. Announcing the event on the other side of the world, US commanders said that "crew members report good results." Ground Zero turned out to be a tennis court in Urakami, but the precise target was hardly relevant because, in seconds, the bomb destroyed the whole of the north side of Nagasaki. Some 40,000 people were killed by the blast and burns from the radiation and firestorms. Later, another 40,000 died from radiation sickness and cancers. The bomb was the second and last atomic weapon ever detonated as an act of war. The first had been dropped on Hiroshima three days earlier.

HIROSHIMA Japan

1945 Hiroshima was slightly bigger than its atomic twin, Nagasaki, with a population of around 240,000. US commanders chose the city ahead of another shortlisted target, the historic city of Kyoto, because one of them had spent his honeymoon in Kyoto. Also, bomb experts felt that the hills surrounding Hiroshima might "focus" the blast and cause more damage. When "Little Boy" detonated, most of the city was destroyed – though some reinforced concrete buildings, such as the tower in the foreground, survived. Almost twice as many died as in Nagasaki – an estimated 140,000 in all. So total was the destruction that all communication with the outside world ceased, and it was 16 hours before the Japanese government learned what had happened – from the White House.

From the ashes Sixty years on, Hiroshima is a bustling, modern city. Its population has recovered to more than a million people, many of them manufacturing Mazda cars for export around the world. Its castle and shrine have been rebuilt and its annual flower festival resumed. Hiroshima has proclaimed itself a city of peace. And its Peace Memorial Park, commemorating the victims of the bomb, is now a popular and poignant visitor attraction.

MEKONG RIVER Vietnam

Late 1960s When the Vietnam War began in the early 1960s, most of
the region's rainforests were intact. Too intact for the American military. The trees
provided perfect cover for their enemies, the Vietcong, who travelled the country
bringing arms to rebels and recruiting villagers. So the US took a leaf out of British
counter-insurgency methods against communists in the Malaysian jungles in the
1950s. They decided to make the Vietcong visible with a virulent weedkiller called
"Agent Orange". Between 1962 and 1971 American aircraft sprayed 50 million litres
of Agent Orange over a fifth of South Vietnam's forest.

Defoliation Agent Orange's ability to defoliate large tracts of rainforest was soon clear across wide areas, as this photo of the Mekong river from 1970 shows. But about then doctors began reporting an epidemic of birth defects among Vietnamese villagers. It emerged that Agent Orange contained a contaminant called dioxin that caused foetal deformities and cancers. The spraying stopped. Today the forests are recovering – where they have not been cut down by loggers. But Vietnamese doctors report high rates of birth deformities even among the grandchildren of those exposed to Agent Orange.

MESOPOTAMIAN MARSHES Iraq

1973 The Mesopotamian Marshes in southern Iraq have for millennia been the largest expanse of wetland in the Middle East, covering an area of around 15,000sq km (5,800sq miles) in the flood season. They have sustained millions of birds and helped maintain fisheries in the Persian Gulf. They have also been home to the Madan people, whose ancient ways of hunting and fishing among the reed beds and waterways lasted for 5,000 years and were chronicled in Wilfred Thessiger's book *The Marsh Arabs*. Some say these marshes were the inspiration for the Garden of Eden myth.

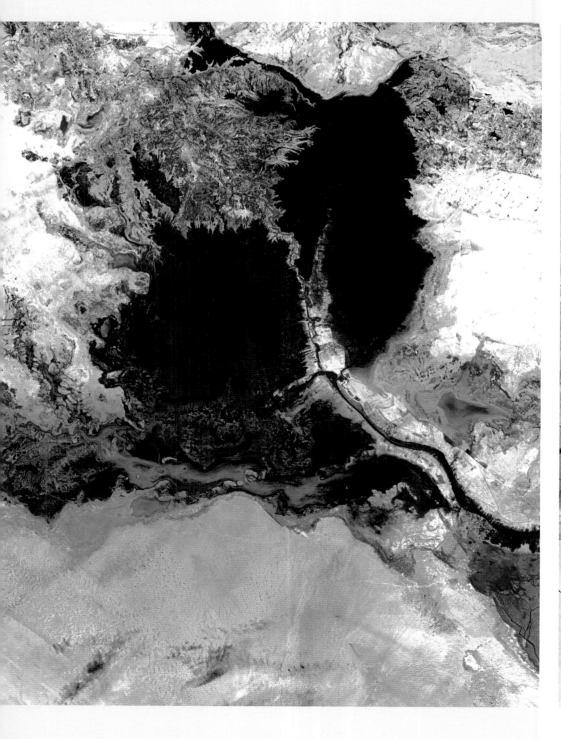

The great drain
Unfortunately, in the 1990s, following an old British colonial blueprint, Iraqi President Saddam Hussein constructed a huge system of dykes and canals to drain the marshes and divert the flows of the two rivers that fed them, the Tigris and Euphrates. The British aim had been to take water to irrigate fields in the desert; Saddam's was to flush out the Madan, who had rebelled against him at the end of the first Gulf War in 1991. Saddam's engineering project, inadvertently helped by dam-building upstream in Turkey, successfully drained most of the marshes before the decade was out, leaving behind massive salt pans and dead vegetation. The Madan fled to Iran.

Reversal
After US and British forces removed Saddam, the Madan began to return, dynamiting the dykes and setting about the task of refilling the marshes. Within a few months large areas had reflooded and by late 2006 more than half of the former marshes had some water. But the water is much saltier than before, so freshwater ecosystems have not recovered and the Madan can no longer rely on the wetlands for drinking water.

BRANDENBURG GATE, BERLIN Germany

1988 Winston Churchill called it an Iron Curtain. The divide between East and West that opened up after the end of World War II was at first ideological, then economic, and, finally, a physical barrier separating the two halves of Europe. At its heart was the Berlin Wall, which ran through the divided city, and the most hated piece of that wall was the barricade constructed by the East German government in 1961 across the Brandenburg Gate, the only surviving ancient gateway into the German capital. The wall's purpose was to prevent the inhabitants of East Germany from fleeing to the freer and more prosperous West. Hundreds were shot in their efforts to climb the wall and make their escape.

Removing the barrier Just as the barricaded Brandenburg Gate became a symbol of Soviet oppression, so the dismantling of the barricade became a symbol of the reunification of Europe. As the East German government fell in late 1989, one of its last acts was to allow its citizens to leave, which they celebrated by tearing down the wall, stone by stone, starting at the Gate. This picture was taken on 3 October 1990, when East and West Germany were formally reunited. Work to clean and refurbish the Gate began in 2000.

MOSTAR BRIDGE Bosnia

1955 This was once the largest single-arch stone bridge in the world. It was 30m (100ft) long, 20m (65ft) high, and made of 456 blocks of stone. When photographed in 1955, the bridge over the River Neretva at Mostar in Bosnia was already more than 250 years old. It had been built during the reign of the Ottoman Sultan Suleiman The Magnificent and joined the Croat and Serbian Muslim sides of the town. The bridge inspired many. Tourists came just to see it, and local men proved their bravery to their brides before marriage by jumping off it.

Destruction War came to Bosnia after the collapse of the state of Yugoslavia in 1991. As battles raged Croat tanks targeted the Mostar Bridge, and in November 1993, just a couple of months after this picture was taken, they destroyed it altogether. The loss of the bridge symbolically divided the town once again, although most Serbs had already fled. The ancient bridge was replaced by a rickety footbridge that served for more than a decade – a poignant metaphor for a fractured country.

Reconstruction After hostilities ended in 1997, Hungarian divers went in search of the old stones on the river bed and Turkish engineers began reconstruction of the bridge at an eventual cost of $13 million. The Mostar Bridge, complete with reinforced foundations, reopened in July 2004 amid great fanfare, as a symbol of a new beginning for the town and the country. But, sadly, few Serbs have so far returned to Mostar.

BEIRUT Lebanon

1982 Lebanon suffered a bloody civil war from 1975 to 1990. Internal factions and external armies battled for control of the country. Parts were occupied by Israelis and Syrians. A further destabilizing influence was the large refugee camps containing Palestinians displaced from their homeland to the south. The capital, Beirut, was heavily bombed by Israeli forces in 1982 as they attempted to root out the Palestinian Liberation Front, controlled by Yasser Arafat. In the process, some 7,000 people died. A brittle peace took hold in 1990, after which Israel withdrew from the country. Syrian forces effectively took over until they, too, withdrew in 2005 after Lebanese street protests against them.

Bombers return Beirut had been extensively rebuilt by the time carnage returned in July and August 2006, as Israel again bombed southern suburbs of the city. This time Israel was attempting to neutralize Hezbollah, a Shia Muslim group that was both a partner in the national government and responsible for launching rockets from southern Lebanon onto villages in northern Israel. Many citizens fled during the month-long barrage, before a ceasefire was agreed. The conflict killed 1,400 people, mostly Lebanese civilians in Beirut. At the end, nothing seemed to have been resolved.

QALQILYA Palestine

2002 Qalqilya is a Palestinian market town on the West Bank. It is close to the pre-1967 border with Israel, and to the trans-Israel highway, which can be seen in the photo on the far left. Its inhabitants have lived for centuries by farming and, in recent decades, by providing labour for Israeli employers. The town also became a major centre for Palestinian refugees who lost their homes and land after the wars with the Israelis in 1948 and 1967. It has a population of some 40,000.

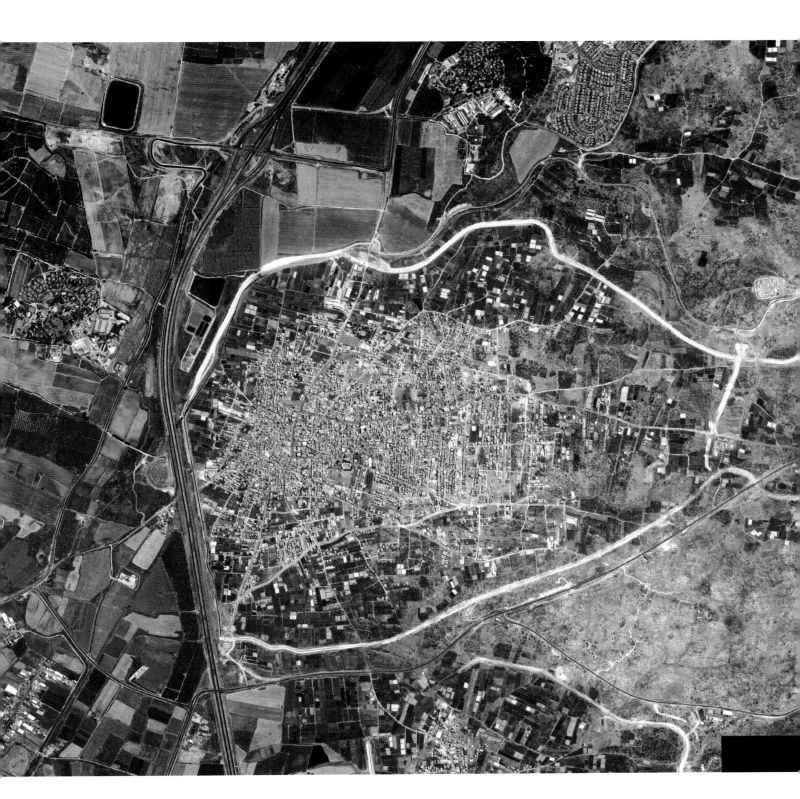

The wall In late 2002, after a Palestinian uprising that killed hundreds of Israelis, the Israeli authorities constructed a "security barrier" to separate Palestinian communities on the West Bank from their Israeli neighbours. Among the first places to be cordoned off was Qalqilya. In this satellite image, taken in June 2003, the barrier is seen as a white line virtually surrounding the town. It comprises an 8m- (26ft-) high concrete wall, a 4m- (13ft-) deep ditch, watchtowers, and an access road. Since its construction outsiders can reach Qalqilya only through an Israeli checkpoint. Farmers are cut off from their fields, workers cannot travel to work, and economic life in the town has come close to a standstill.

TWIN TOWERS, NEW YORK USA

Early 1990s Ever since their completion in 1972, the Twin Towers of the World Trade Center in New York were a symbol of successful global capitalism. Each was 110 storeys high, the tallest buildings in the world. Some 50,000 people worked in them every day, mostly for major financial companies such as Morgan Stanley. The centre had a troubled history: it was hit by fire in 1975 and a truck loaded with explosives was detonated in the underground car park of the North Tower in 1993. Islamic extremists later convicted of the bombing said their aim had been to destroy both towers.

Terrorism Despite this warning, nothing prepared New York – or the world – for the destruction of the Twin Towers on 11 September 2001. Terrorists from Al Qaida flew packed passenger aircraft into each tower in turn, within 17 minutes of each other. Both towers collapsed in the hours that followed, with the loss of some 2,700 lives. The debris smouldered for 99 days. Ground Zero, where the two towers once stood, remains a hole in the heart of the world's premier financial district. The US subsequently invaded Afghanistan and toppled the Taliban, whose leaders had harboured Al Qaida. But, at time of press, the USA has failed to track down the masterminds of the assault on the Twin Towers.

Leisure and Culture

What do we do when we are not making war? We make peace, we make art – or we go on holiday. Leisure reveals the human imagination untrammelled by practical needs. But it is big business, too. The annual holiday is a ritual for most of the world. And if you can, you have two or three or four. The skies and roads are full of people seeking inspiration and pleasure by going somewhere fresh to see or hear or experience something new.

Sometimes, admittedly, our demands are for basic sensory pleasures in a mildly exotic location. An area such as the Mediterranean – the world's top holiday destination. A firm favourite there is Benidorm, a Spanish fishing port whose white sandy beach has packed in so many holidaying Europeans in search of sun, sand, and sangria that it has become the "Manhattan of the Mediterranean". Or somewhere like the world's largest golf resort, created for Chinese businessmen on a construction site in Guandong.

More often than not we want spectacular scenery too. Cue the growth of Alpine resorts such as Zugspitze, once no more than a wind-blasted Bavarian meteorological observatory, but one that offered spectacular views across four countries. Then there is Zermatt, the exclusive car-less Swiss resort beneath the Matterhorn.

A sense of history goes down well too. "Heritage" as the tourist trade now calls it. Mount Rushmore attracts millions of Americans who queue to see the images of four famous presidents dynamited into the granite peaks of South Dakota. Bizarrely, so too does London Bridge, which was relocated from its River Thames home to a canal in Arizona and became one of the state's top tourist attractions.

But Londoners bemused by the attractions of a hand-me-down bridge have their own transported piece of history. Cleopatra's Needle, an Egyptian obelisk that spent the best part of 2,000 years beneath the sands of Alexandria in Egypt before being relocated to the banks of the Thames, is on most tourist routes through the city.

History, many would agree, is better when not transported. The Buddhas hewn from sandstone cliffs across Afghanistan qualified as true examples of cultural heritage. But that was before their desecration at the hands of the Taliban. Other genuine cultural treasures include the Mayan temples in the jungles of Central America and Luxor in Egypt. The latter, like Cleopatra's needle, was rescued from the Saharan sands but, unlike the incongruous obelisk, has been left in its proper setting.

Context can be lost, of course. Often we end up despoiling what we come to see. The Great Wall of China is one of the great monuments of human history – famously visible from the moon. But the ancient barrier of the Middle Kingdom against Mongolian hordes is now corralled by the highways that transport the millions who come to see it.

Sometimes, however, tourist sites can even gain from their incongruous settings. The Eden Project, one of Britain's most popular attractions, is a celebration of the world's ecosystems and the wealth that mankind has created from them. The effect is enhanced by the fact that the site is an abandoned china-clay pit in Cornwall. And what price a sandy beach in the middle of Paris? For peace, art, *and* a holiday, maybe nothing quite had the allure of the hippie nirvana that was the Woodstock festival. But that's history.

Wherever we go, we leave litter. Nowhere is safe. Not even here, high up in the Himalayas. The ridge of mountains known as Annapurna includes the 10th-highest peak in the world. These are among the most dangerous mountains to climb, and deaths are frequent. But even so, trekkers get here and leave their rubbish. It's got so bad that some mountaineers now take up teams of "eco-climbers", who are dedicated to removing the rubbish.

MOUNT RUSHMORE USA

1925 The Black Hills of South Dakota are imposing. The thrusting granite peaks are said to be America's oldest mountains, dating back at least 60 million years. The surrounding pine and spruce forests were once the domain of the Lacota Sioux people. But, like the state itself, they were little visited by outsiders until Doane Robinson, lawyer and state historian, had the idea of creating giant carvings of national heroes on the exposed rock-face of Mount Rushmore as a means of attracting tourists.

Reprofiling The first carving on Mount Rushmore began in 1927. But it took some 400 men, headed by the chief sculptor Gutzon Borglum, 14 years to complete the task. They removed half a million tonnes of rock, mostly by dynamite, to create the 20m- (65ft-) high faces of four of the US's most esteemed Presidents: George Washington, who led the War of Independence against Britain; Thomas Jefferson, author of the Declaration of Independence; Theodore Roosevelt, revered early-20th-century President; and Abraham Lincoln, who campaigned to abolish slavery. Some three million visitors come each year as an act of reverence for their nation. The hard granite rock should ensure the survival of the faces: erosion rates are estimated at just 1cm (³⁄₈in) every 400 years.

LUXOR TEMPLE Egypt

1857 For almost four millennia, two statues of Ramesses II, one of ancient Egypt's greatest kings, have stood at the gateway to the temple at Luxor in Egypt. The temple was the heart of the great city Thebes at the height of ancient Egyptian civilization. In this Victorian picture, the statues appear to be head-and-shoulders images. But look more closely. They are surrounded by piles of sand. The Sahara desert had been invading the temple for centuries, filling its rooms and piling up against its walls. There is more to these statues than meets the eye.

Clean sweep Thanks to assiduous restoration and the removal of several metres of sand from the temple floor, the statues are today revealed in their former glory. Far from being head-and-shoulder images, they are giant carvings of a seated Ramesses, depicting everything down to his toes. Luxor, on the east bank of the Nile in Egypt, is today one of the major tourist attractions of the world. It is visited by millions every year. Keeping Luxor free of invading sand is a constant job, as the sweeper at work in the foreground shows. But have restorers reached the true floor of the temple? Maybe not. The temple itself is built on a slight rise, and some believe that beneath it, still unexcavated in the sand, lie a previous temple and maybe more statues.

BAMIYAN VALLEY Afghanistan

Desecrated idol In February 2001, Mullah Mohammad Omar, the Supreme Commander of the Taliban, ordered the destruction of all statues in the country. Islamic teaching, he insisted, abhorred idols of all sorts, whether cultural treasures or not. Armed with dynamite and mortars, his troops spread out across the country to fulfil the order. And on 8 March the Bamiyan Buddha was reduced to rubble, leaving behind a bare gap in the cliff where the statue had stood for almost two millennia.

1987 Some 125km (80 miles) west of Kabul, in the heart of the Hindu Kush mountains, stood the world's tallest statue. At 53m (175ft) high, this representation of Buddha, hewn from the sandstone cliffs of the Bamiyan Valley, counted as one of the world's greatest cultural treasures. UNESCO had designated it and a smaller companion statue as a World Heritage Site. The statues had been carved by the people of the ancient Buddhist kingdom of Kushan between the second and fifth centuries AD, when the caves in the cliffs housed a convent. The convent was abandoned after Islam came to the region. But though neglected and occasionally attacked thereafter, the statues survived for more than 1,000 years. Until the Taliban took power in Afghanistan.

MAYAN TEMPLE Central America

1955 Tikal is the largest of the "lost" cities of the Mayan people, who once flourished in the jungles of Central America. When first surveyed in the 1950s, many of the temples of Tikal were tall enough, at around 60m (200ft), to penetrate the tree canopy. But the huge pyramids on which they stood were entirely covered by vegetation. Tikal was a major metropolis for almost 1,500 years, between the fourth century BC and the 10th century AD. The city spread across 60sq km (25sq miles). Besides temples and palaces, it built universities and giant irrigation schemes that watered farms cleared from the jungle. At its peak, Tikal had a population of up to 200,000.

Re-emergence After centuries forgotten in the Guatemala jungle, the main palaces and temples of central Tikal – like this one, known simply as Temple 1 – are now a major tourist attraction. But even today, after half a century of effort by archaeologists, only a fraction of the ancient city has been surveyed. Many lesser buildings lie shrouded in vegetation. One secret is becoming clear, however. The Mayan civilization seems to have disappeared because of climate change. A long drought destroyed its farming systems. The people in the cities had no food – and eventually disappeared into the jungle to forage for survival. The "primitive" Indians discovered by Europeans centuries later were the survivors of a great civilization.

CLEOPATRA'S NEEDLE Egypt/England

c1840 Cleopatra's Needle was erected 3,500 years ago in Heliopolis by an Egyptian pharaoh called Tuthmosis III. It was one of three, each 21m (69ft) high and made of solid granite hewn from a quarry near Aswan on the Nile. They were later moved to a temple in Alexandria that had been erected by Cleopatra – hence the name. But the obelisks then disappeared from history, buried in sand, for 2,000 years before archaeologists recovered them. One was given to Britain in 1819 to celebrate victories in the naval battles of the Nile and Alexandria. But the British government refused to fund its move to London, so it stayed for several decades behind a wooden fence on the outskirts of Alexandria, where this picture was taken.

Transportation In 1878 a public subscription finally raised £15,000 to pay to bring the obelisk to Britain. It set out in a specially made cigar-shaped iron pontoon, towed by a steam ship. In the Bay of Biscay the pontoon capsized in a storm and six crew drowned. But the partly swamped vessel was recovered five days later and eventually sailed up the River Thames more than three months after setting out. It was raised on the Thames embankment, where it has sat ever since – a small piece of ancient Egypt on the streets of London. Two other "Cleopatra's Needles" also found their way to foreign parts in relatively recent times: one is in Paris and the other in New York.

LONDON BRIDGE England/USA

1895 London Bridge spans the River Thames between the British capital's financial centre, the City of London, and Southwark on the south bank. The site has had a bridge since Roman times and until as late as 1750 it was the only bridge over the river in London. Old wooden bridges repeatedly burned down, but when stone versions were built in the Middle Ages they were colonized on top by shops and houses, and beneath by water wheels to grind corn. This version, made of Dartmoor granite, was completed in 1831 by the famous engineer John Rennie and was free of buildings. Even so, it began to sink into the Thames mud and was replaced in the 1960s.

Transatlantic trip In 1968 Rennie's bridge was sold to US oil magnate Robert McCulloch for $2.4 million. He dismantled it stone by stone and shipped it across the Atlantic for reassembly (though not all of it: some stone was impounded in the UK in lieu of taxes). Some say that McCulloch believed he was buying the rather more distinctive Tower Bridge nearby. But it proved a smart purchase, anyhow. Today the old stone forms the cladding for a concrete bridge that spans a canal at Lake Havasu City in Arizona. It is the state's second-most popular tourist attraction, exceeded only by the Grand Canyon.

GREAT WALL China

1871 Once, the Great Wall of China was the farthest reach of the greatest empire on Earth: an impregnable barrier between the Middle Kingdom and the barbarian hordes in Mongola and beyond. The wall was constructed in stages from the fifth century BC. It is by some way mankind's longest structure, snaking along mountain ridges for some 6,400km (4,000 miles), and regularly dotted with forts and armed border posts. Even during the 20th century, many parts of the wall were in an empty landscape, such as this section in the Shuiguan Valley not very far from Beijing.

Change and decay This is the same spot in the Shuiguan Valley today. The wall's location has been wrecked by a major highway and tunnel. The Great Wall now appears almost mundane in the face of modern civil engineering. A survey conducted in 2003 found that a third of the Great Wall had disappeared, either knocked down for roads and other developments, or scavenged for stone. Another third was badly damaged. The wall is becoming a victim of the greatest building boom ever seen on the planet. A fitting riposte to the Great Wall, some might say. But will today's monuments also be revered in two millennia?

QINGHAI–TIBET RAILWAY Tibet

1987 In the mountains of Tibet, yaks are the most common animals. And yak hide is the material of choice for constructing small boats to cross the streams that drain from the glaciers of the great Tibetan plateau. Here, in the middle of the long dry season between monsoons, when the rivers dwindle, two Tibetans are using their yak-hide boats to cross the River Lhasa near the Tibetan capital of the same name. The image could have come from almost any era – certainly from a time when Tibet was an independent country. But since 1951 it has been under the control of China.

Crossing rivers And this is how the Chinese like to cross rivers. In 2006, the new Qinghai–Tibet railway opened. It is the world's highest railway; at one point it climbs to more than 5,000m (16,000ft) above sea level. More than 1,100km (700 miles) long and built largely across permafrost, it is designed to bind restive Tibet ever closer to Beijing. Here it is crossing the River Lhasa over the new La Sa Te bridge, near its terminus in Lhasa.

WOODSTOCK USA

1969 "We are stardust; we are golden; and we've got to get ourselves back to the garden." So sang Joni Mitchell, troubadour of the half a million hippies and counter-culture idealists who thronged to 250 hectares (600 acres) of cow pasture near Bethel in upstate New York one rainy weekend in mid-August 1969. They were attending the most totemic of all the outdoor rock festivals – Woodstock. Jimi Hendrix, The Who, Santana, Joan Baez, Janis Joplin, and The Grateful Dead all played on Max Yasgur's dairy farm. In keeping with the hippie culture it was free – albeit only after the crowds had torn down the entrance gates.

A different tune There were a few attempts to recreate the Woodstock festival in subsequent years, but its "karma" was a one-off. And the "peace-and-love" hippie movement was soon splintered by hard drugs, insanity, money-making, and radical politics. But music has not entirely deserted these pastures. The photo shows the same site 37 years later. It has been bought by a new generation of arts entrepreneurs and 2006 marked the opening, just over the hill, of the $70-million Bethel Woods Arts Center dedicated to – classical music.

ZERMATT Switzerland

1900 The Swiss village Zermatt sits at the head of a valley beneath the Matterhorn, the most famous mountain in the Alps. The village was once occupied exclusively by migrant pastoralists. But Zermatt's fate has been bound to that of the mountain since the peak was first climbed from there by a British team in 1865 – a conquest made all the more famous because four of the seven mountaineers fell to their deaths as they returned to Zermatt. This image was taken in about 1900, shortly after the completion of a railway up the valley. It shows early tourists visiting chalets in the village. The railway brought thousands of visitors every summer. In the days before skiing, most people just came to stand and stare.

Controlled growth Zermatt is much bigger today, with a permanent population of more than 5,000, and more than 13,000 visitors' beds. The chalets have become small apartment blocks. But with no road access, it remains a village without cars. Electric buggies and horse-drawn sleighs are the only transport allowed, though the air is sometimes filled with the sound of an arrival at the heliport, where four helicopters will take well-heeled visitors for a tour around nine of the ten highest mountains in Europe.

ZUGSPITZE, BAVARIA Germany

1925 Zugspitze, in the Bavarian Alps right on the border with Austria, is the highest mountain in Germany. The mountain was not climbed until 1820, but in 1900 this observatory was built on its summit to provide weather reports for both Germany and Austria. At almost 3,000m (10,000ft), it was a remote and forbidding outpost that could only be reached by a two-day trek through the Partnachklamm Gorge.

Access for all The scene today is rather jollier. The wind-blasted outpost of meteorology has become a tourist attraction, providing stupendous views across four countries and some of the best skiing in the Alps. Access is easy these days. The summit of Zugspitze is reached by three different cable cars, two from Germany and one from Austria. And a cog railway improbably tunnels its way to within 400m (1,300ft) of the summit. Early risers get breakfast at the top, and the observatory even houses its own border checkpoint.

CORNWALL England

1999 There can be few more unprepossessing sights than the sterile depths of a disused china-clay pit. The clay pits of Cornwall in southwest England have been in decline for many years, leaving behind huge, soil-less, waterlogged holes in the ground as well as vast piles of white spoil, and making parts of Cornwall into a virtual moonscape. But this pit, at Bodelva just north of St Austell, attracted the attention of Tim Smit, a former record producer turned aficionado of gardens. Improbably, he wanted to build in it the world's largest conservatory, to showcase plants and how we use them.

Eco-tourism Thanks to a large national lottery grant and a lot of favours, this
is the result – the Eden Project. The pit has become a vast terraced garden with, as its
centrepiece, a series of geodesic domes. Inside, Smit has recreated entire ecosystems,
such as rainforests and the Mediterranean, and displays teach visitors how the plants
are used to provide food and clothing, and make fuel, building materials, and medicines.
The largest conservatory, containing a huge array of giant tropical trees and plants that
almost burst out of the dome, is 240m (790ft) long and 50m (165ft) high. Since it opened
in 2001, some seven million visitors have descended into this Cornish Eden.

SHENZHEN China

2003 It's pretty much all bunker in this shot. Jose Maria Olazabal, one of the world's top golfers, takes a tee shot in the midst of a construction site somewhere between the two electronics boom towns of Shenzhen and Dongguan in southern China. Why?

Leisure landscaping The first image was a publicity shot, of course. It was designed to show the transformation of the landscape during the eight months between that shot and this. The same golfer in the same spot now has no need of a tee to play this shot over the water to the green on the par-five 15th at the Missions Hills Golf Club. Olazabal designed this particular course, but it is one of 12 18-hole courses at what is claimed to be the world's largest golf resort, "only 20 minutes from Hong Kong." Golf is booming all across Asia, as the newly rich indulge in the sport of the Western well-to-do. And China, as in most things, is racing hard to catch up.

DAL LAKE India

1986 This Himalayan idyll is the Dal Lake. Surrounded by the magnificent mountains of Kashmir on India's border with Pakistan, the area has in the past been sustained by a thriving tourist industry. Many visitors stayed in houseboats on the lake. Its shores were covered in gardens and orchards, the produce from which was sold in floating vegetable markets like this one on the lake itself.

Turning their backs But the escalating conflict between India and Pakistan over the status of Kashmir has frightened away most of the tourists. Deprived of any incentive to retain the lake's beauty, local residents have begun building houses on its shore. The illegal construction has upset the hydrology, and effluent from the townships has turned parts of the lake into a quagmire of floating weeds. Having once been the hub of local life, the lake is increasingly an abandoned eyesore. Residents have turned their backs on its waters. The markets have closed – and even the old trading boats are sunk.

PARIS France

1975 Notre Dame Bridge over the River Seine is in the heart of Paris. Either side of the river, busy highways keep the city moving while tourist boats ply the river itself. Beyond the bridge, on the Ile de la Cité, is the giant Palais de Justice, the centre of the French legal system and a former palace.

Seasonal sands But every summer since 2004, Paris has forsaken its riverside highways to turn the right river bank into the "Paris Plage". More than 2,000 tonnes of sand covers the tarmac; there are bandstands, and palm trees, and sunloungers, and marquees, and even a swimming pool. Millions of tourists are joined by Parisians who once deserted the city during August, but now stay to enjoy the fun.

JUMEIRAH BEACH, DUBAI United Arab Emirates

2003 What do oil sheiks do with their spare cash? The answer, increasingly, is that they put it into real estate. Not just buying old mansions and castles in Europe, but creating new real estate in the deserts, and even the seas, of the Middle East. Dubai, one of the United Arab Emirates in the Persian Gulf, is among the fastest-growing cities in the world. But besides turning into a great trading hub, the emirate is also becoming a holiday destination for the rich. In this satellite image, Dubai itself is towards the top right, centred on its creek. But stretching south along the coast is the Jumeirah Beach. And off the beach stretches the world's largest man-made island. Constructed in the shape of a palm tree, it is called The Palm.

Growing palms Three years on, the "palms" are multiplying. These complexes of private holiday islands are favoured by, among others, English football stars. The original complex, now known as Palm Jumeirah, has been joined by the larger Palm Jebel Ali to the south. And in this 2006 image, construction has just begun on the still-larger Palm Deira. It will eventually comprise some 8,000 houses on 250 small islands, and overall will be 14km (9 miles) long and 8km (5 miles) across.

BENIDORM Spain

1963 Back in the 1960s, when most Europeans still holidayed in their home country, Benidorm was a small fishing village near Alicante, on the Mediterranean coast of Spain. Its domed church on a rocky peninsula, from where this picture was taken, provided good views of a particularly fine 6km (4-mile) stretch of white, sandy beach in three bays. But with travel companies set to start organizing cheap package holidays to the "Costa Blanca", its seclusion was about to end.

High-impact growth Today in Europe only London and Milan have more high-rise blocks than Benidorm. The "Manhattan of the Mediterranean" has 300 buildings exceeding 35m (115ft) in height. It is said to have more hotel "stars" than the whole of Greece. Benidorm's permanent population of 65,000 grows tenfold during the holiday season when, despite the attractions of budget flights to an increasing number of cities across Europe, the Spanish coastal resort maintains its place among the Continent's most popular holiday destinations. More than a million British visitors alone come here to sample its casinos and discotheques each summer.

INDEX

A

Aberdeen, Earl of 71
Aceh, Sumatra 192–5
Afghanistan 247, 253
Africa 10, 12, 34, 50–1
agriculture see farming
Air Nauru 151
air pollution 119, 158–9, 176
air travel 17
Al Qaida 245
Alaska 182–3
Albatross Bay, Queensland 148–9
Albert of Agostini, Mar'a 30
Alexandria 247, 256
Almeria 134
aluminium 119, 148–9
Amazon jungle 12
Amazon river 136–7
Anatolia 124–7
Anchorage 182–3
Andes, Peruvian 24, 34, 163, 184–5
Annapurna 247
Antarctica
    Antarctic peninsula 38
    Filchner ice shelf 36–7
    hole in ozone layer 65
    Larsen B ice shelf 38
Aral Sea 10, 11, 15, 54–5, 119
Arctic Ocean 21, 40–1
Argentina 30, 86–7, 140–1
Argentino Lake 30
Arizona, London Bridge 247, 258–9
Atacama desert 15, 146–7, 206–7
Ataturk Dam, Turkey 124–5, 126
atomic bombs see nuclear weapons
Australia 102–3, 148–9, 190–1
avalanches 163, 178–9 see also landslides

B

Baez, Joan 264
Bam, Iran 163, 180–1
Bamiyan Valley 253
bauxite mining 119, 148–9
Bavaria 268–9
Beaumont-Hamel 212–13
Beijing 67, 96–7
    Tiananmen Square 62–3

Beirut 209, 240–1
Belgium 210–11
Benidorm 247, 280–1
Berlin, Brandenburg Gate 209, 236
Berlin Wall 236
Betsiboka River 200–1
BHP Billiton 146
Bilbao 15, 67, 112
Bingham Canyon 15, 119, 144–5
Bolivia 82–3
Borglum, Gutzon 249
Bosnia 238
Brazil 136–41 see also São Paulo
bridges 88–91, 238, 258–9, 263
Buddha statues 247, 253
Buenos Aires 67
    Retiro shanty town 86–7

C

Caen 209, 216
Campo de Dallas, Spain 134
Canada 152
Canadian Arctic 40–1
Canadian forces 216
    Newfoundland regiment 212
Cape York, Queensland 148–9
carbon emissions 15, 16
Carysford, HMS 58
Chad 21, 50–1
Charlie, Bonny Prince 88
chemical pollution 67, 74, 119
        see also ozone layer, hole in
Cheonggyecheon river 108–9
Chernobyl 19, 156–7
Chile 146–7, 206–7
China 21, 46–7, 97, 272–3 see also Beijing; Hong
        Kong; Tibet; Yangtze River
    Great Wall 247, 260–1
    Lake Dongting 42–3
Chinese migrants to US 78
Chongqing 21, 122, 123
Churchill, Winston 236
cities 2, 18, 44–5, 52–3, 62–3, 67–87, 96–103, 106–17
Ciudad del Este 140, 141
Cleopatra's Needle 247, 256
climate change 17, 21, 163 see also glaciers,
        melting; ice sheets, disintegrating; ozone
        layer, hole in
CNN 49

coastal erosion 180–91
Cold War 209, 236
Cologne 18
    Hohenzollern Bridge 18, 218–19
Colorado River 119, 120
conflict 16, 18, 209, 234–45 *see also* war
Copenhagen 92
copper mining 15, 119, 144–7
coral reefs 56–9
Cordillera Blanca mountain range 24
Cornwall 247, 270–1
cotton production 10, 54–5, 126–7
Croats 238
Culloden, battle of (1746) 88
Cultural Revolution 62
culture and leisure 247–81
Cyclone Gafilo 200–1

D
Dal Lake, India 274–5
Dammastock Peak 28–9
dams 114, 119–27, 141
defoliation 209, 232–3
deforestation 10, 12, 13, 16, 21, 42–3, 60–1, 82–3,
    129, 132–3, 136–41, 142, 161, 163
deltas, silting of 21, 46–7, 200–1
Denmark 92
Depression, Great 80
deserts, irrigation of 12, 130–1
Dongguan 247, 272–3
Dresden 209
    Frauenkirche 15, 220
droughts 21, 204–5
Dubai 2, 67, 72–3, 278–9
Dusseldorf 52–3
dust storms 15, 21, 55, 204–5 *see also*
    sandstorms

E
earthquakes 76, 78–9, 84, 163, 180–7, 182–3
Ecuador 132–3
Eden Project 247, 270–1
Edwards, Mark 87
Egypt 250, 256
Eiger 33
El Niño 61, 161, 207
Elbe, River 21, 48–9
Eldfell 166–7
Elugelab island 209

England 106–7, 188–9, 214–15, 256–8, 270–1
environmental change 21
environmental rehabilitation projects 15
Escondida Mine 119, 146–7
Euphrates, River 119, 124–5, 126, 235

F
farming 10, 12, 13, 21, 119, 134
Felixstowe 106
Filchner, Wilhelm 36
Finland 119
fish farming 15, 132–3, 137
fish stocks, reduction of 13, 15, 132
flooding 21, 42–9, 163, 200–3 *see also* sea levels,
    rising; tsunami, Indian Ocean
Florida Keys, Carysfort Reef 58–9
forest fires 60–1, 161
forestry 119, 142 *see also* deforestation
Foster, Sir Norman 90
France 90, 212–13, 216, 276–7

G
Germany 18–19, 48–9, 52–3, 209, 218–20, 236,
    268–9
Gibraltar 67, 104
Gifford Pinchon National Forest 142
glaciers, melting 16, 21, 22–33
Glen Canyon 120
Gletsch 28–9
global warming 17, 21, 163 *see also* glaciers,
    melting; ice sheets, disintegrating; ozone
    layer, hole in
golf 272–3
Grapes of Wrath, The 204
Grateful Dead, The 264
Grimsel Pass 28–9
Grinnell, George Bird 26
Grinnell Glacier 26–7
Guandong 247, 272–3
Guatemala 254–5
Guggenheim Museum, Bilbao 112
Gulf of Guayaquil 132–3

H
Hawaii 172–3
Heimaey Island 166–7
Hendrix, Jimi 264
herbicide, Agent Orange 209, 232–3
Hezbollah 241

Himalayas 163, 186–7, 247, 274–5
Hiroshima 16, 209, 229, 230
Hitzacker 48–9
Homo sapiens 10, 163
Hong Kong 68
    Aberdeen Harbour 71
Hoover Dam 114, 120
human activity, recovery from 12–13, 19, 157
Hurricane Katrina 163, 196–9, 203
Hussein, Saddam 209, 235

I

ice sheets, disintegrating 15, 16, 34–41
icebergs 37, 38
Iceland, islands 164–7
Iguazu National Park 140
India 274–5
Iran 180–1
Iraq 234–5
irrigation 12, 15, 124–7, 130–1
Isahaya Bay, Japan 154–5
Israel, security barrier 209, 243
Israeli forces 240, 241
Itaipu Dam 141
Italy 44–5, 158–9

J

Jacabamba glacier 24
Japan 154–5, 163, 228–30 *see also* Tokyo
Jefferson, Thomas 249
Jialing, river 21
John Paul II, Pope 159
Johnston, David 169
Joplin, Janis 264
Jordan 12
Jordanians 131
Jumeirah Beach, Dubai 278–9

K

Kashmir 186–7, 274, 275
Kathmandu 67, 101
Kazakhstan 55 *see also* Aral Sea
Kiribati 224–5
Kobe 163
Korea, South 108–9
Kowloon Peninsula 68–9
Krkonose Mountains 49
Kuala Lumpur 160–1
Kyle of Lochalsh 88

L

Lake Chad 21, 50–1
Lake Dongting 42–3
Lake Geneva 28, 29
Lake Havasu City 259
Lake Powell 120
Lake Saimaa 119
land reclamation 154–5
land transformation 119–61
landslides 163, 185–7 *see also* avalanches
Las Vegas 67, 114–15
Lawrence of Arabia 12
Lebanon 240–1
leisure and culture 247–81
Leningrad (now St Petersburg)
    Nevski Prospekt 222–3
    siege of 16, 209, 222
Lhasa, River 262–3
Lincoln, Abraham 249
London 67
    Canary Wharf 67, 106–7
    Cannon Street 214–15
    Cleopatra's Needle 247, 256
    Isle of Dogs 106–7
    London Bridge 258
    Millennium Dome 107
    Tower Bridge 214–15
    West India Docks 67, 106–7
London Bridge (Arizona) 247, 258–9
Los Angeles 67, 109, 117
Luxor 247, 250
Lynas, Bryan and Mark 24

M

Machala, Ecuador 132–3
Madagascar 200–1
Madan people (Marsh Arabs) 209, 234, 235
Makhri, Pakistan 186–7
Malaysia 160–1
Maldives 21, 56–7
mammals, large 13
mangrove forests 15, 132–3, 136–7
Mao Ze Dong 62, 96, 97
Marsh Arabs (Madan people) 209, 234, 235
Marsh Arabs, The 234
marshes, draining of 13, 16, 163, 202, 234–5
Matterhorn 247, 266–7
Mayan temples 12, 247, 254–5
McCulloch, Robert 259

Mekong River 232–3
Merkel, Angela 49
Mesopotamian marshes 209, 234–5
Mexico City 67, 74–5
Millau Viaduct 15, 90
mining 15, 16, 119, 144–52
Missouri River 202–3
Mitchell, Joni 264
Mohammed Omar, Mullah 253
Montana, Glacier National Park 26–7
Montserrat 163, 174–5
Mostar Bridge 15, 209, 238
Mount Cook 163, 178–9
Mount Huascaran 184–5
Mount Kilauea 172–3
Mount Kenya 34
Mount Kilimanjaro 21, 34
Mount Pinatubo 163, 176
Mount Rushmore 247, 248–9
Mount St Helens 142, 163, 168–71
Mount Toba 163
mountain erosion 33
Muynak 10, 54

N
Nagasaki 16, 209, 228–9, 230
NASA 40
nature, forces of 16, 163–207
Nauru 15, 119, 150–1
Neath, Vale of 94
Neelum, River 186–7
Nepal 101
Neretva, River 238
New Mexico, Trinity site 209, 226–7
New Orleans 163, 196–9, 203
New York 67
    John F Kennedy International Airport 17
    Twin Towers (World Trade Center) 16, 209, 245
New Zealand 178–9
North Pole 40–1
nuclear weapons 16, 209, 226–31

O
Olazabal, Jose Maria 272, 273
ozone layer, hole in 16, 65

P
Pakistan 186–7
Palestine 242–3

Palestinian Liberation Front 240
Palm Jumeirah, Dubai 278–9
Panama Canal 15, 128–9
Para State, Brazil 136
Paraguay 140–1
    Acaray reservoir 141
Parana, River 140–1
Paris beach 15, 247, 276–7
Passchendaele 16, 209, 210–11
Patagonia 30
paving over large areas 15, 16
Peru 24, 184–5 *see also* Andes,
        Peruvian
    Cordillera Blanca mountain range 24
Philippines 176
phosphate mining 15, 119, 150–1
Plymouth, Montserrat 174–5
Pole, North 40–1
pollution 15, 18, 67, 74, 119, 158–9,
        176
Powell, John Wesley 120
prawn farming 15, 132–3, 137
Pripyat, Ukraine 18–19, 156–7

Q
Qalqilya 242–3
Qinghai-Tibet railway line 263

R
railways 263
rainforests 138–41, 161, 209
Ramesses II, king 250
Red Army 10
Rennie, John 258
Rhine, River 21, 52–3, 218–19
Rhône, River 28, 29
Rhône Glacier 28–9
Rio Branco, Brazil 138
Rio Tinto 145, 146
rivers, damming and diverting water from 10,
        15–16, 21, 42–3, 50–5, 119
road-building 15, 94, 260–1
        see also bridges
Robinson, Doane 248
Rome, St Peter's Basilica 158–9
Rondonia region, Brazil 138–9
Roosevelt, Theodore 249
rubbish, leaving 247
Russia 60–1, 222–3

S

Saddam Hussein 209, 235
St Petersburg (formerly Leningrad) 16, 209
    Nevski Prospekt 222–3
Sakhalin Island forest 60–1
San Andreas Fault 76
San Francisco 15, 76–7
    City Hall 78–9
sand storms 63 *see also* dust storms
Santa Cruz de la Serra 82–3
Santana 264
São Paulo 67
    Avenue Paulista 98–9
Sassoon, Siegfried 211
Saudi Arabia, Wadi as Sirhan 130–1
Scarborough, Holbeck Hall 188–9
Scotland 88
sea levels, rising 8, 20, 44–5, 56–7 *see also* flooding
Seattle 67, 80
Seoul 67, 108–9
Serbs 238
Shanghai 109
Shenzhen 272–3
shrimp farming 15, 132–3, 137
Shuiguan Valley, China 260–1
silting of deltas 46–7, 200–1
Singapore 67, 110–11
Skye road bridge 88
Smit, Tim 268, 269
soil conservation 205
Somme, Battle of the 212–13
Soufrière volcano 174–5
South Chad Irrigation Project 51
South Dakota 204–5, 247, 248–9
soybeans 83
Spain 112, 134, 280–1
Sprogo island, Denmark 92
Storebaelt link, Denmark 92
Sumatra 161, 192–5
Surtsey Island 164–5
Swiss Alps 22, 28–9
    Stieregg Restaurant 33
Switzerland 22, 28–9, 33, 266–7
Sydney 102–3
Syrian forces 240

T

Taliban 245, 247, 253
Tanka boat people 71
Tanzania 34
Tarawa, Battle of 224–5
Tarn Gorge 90
terrorism 16, 209, 241, 245
Thames, River 257, 258
Tibet 262–3
tidal waves 163 *see also* tsunami, Boxing Day 2004
Tigris, river 235
Tikal, Central America 254–5
timber production 119, 142, 152 *see also*
    deforestation; forestry
Tokyo 16, 67
    Ginza shopping district 84
Trift glacier 22
Trinitite 226–7
tsunami, Indian Ocean (Boxing Day 2004)
    57, 136, 163, 192–5
Turkey 124–7
Tuvalu 8
Twelve Apostles, Australia 190–1

U

Ukraine 156–7
UNESCO 253
United States of America 17, 26–7, 58–9, 226–7,
    245, 259 *see also* New Orleans; New York;
    San Francisco
    and forces of nature 168–73, 182–3, 202–5
    and land transformation 120, 142–5
    and leisure and culture 248–9, 264–5
    and urbanization 80, 114–17
Upsala Glacier 30
Urakami river 228–9
urbanization 2, 67–117
Urugua-i-Dam 141
US Marines 224
Utrecht, Treaty of (1713) 104
Uzbekistan *see* Aral Sea

V

Vancouver 152
Venice, St Mark's Square 21, 44–5
Vestmannaeyjar 166–7
Vietcong 209, 232
Vietnam 209, 232–3
volcanoes 16, 163–77

W
Wadi as Sirhan, Saudi Arabia 130–1
Wadi Rum, Jordan 12
Wales 94
Wanzhou 15
war 16, 18, 209–33 *see also* conflict
Washington, George 249
Washington state 142
Waters, Thomas 84
Weipa, Australia 119, 148–9
Weyerhaeuser 142
whale species, reduction of 13, 15
Who, The 264
Wiessmies Mountain 22
Woodstock Festival 247, 264–5
World War I 209, 210–13
World War II 18, 209, 214–31

Y
yaks 263
Yangtze River 21, 119
    Lake Dongting 42–3
    Three Gorges Dam 15, 122–3
Yasgur, Max 264
Yellow River and delta 21, 46–7
Yungay, Peru 163, 184–5

Z
Zealand, Denmark 92
Zermatt 247, 266–7
Zugspitze 247, 268–9

ACKNOWLEDGMENTS

Mitchell Beazley would like to acknowledge and thank the following for providing images for use in this book.

Page 2 Getty Images/Gulfimages/Image Solutions/Insy Shah; **6** courtesy Philip's; **8** Still Pictures/Mark Lynas; **11** Getty Images/Robert Harding Picture Library/Tony Waltham; **12** Corbis/Yann Arthus-Bertrand; **14** Corbis/Gilles Sabrié; **17** Corbis/David Jay Zimmerman; **18** Corbis/Bettmann; **20** PA Photos/AP; **22, 23** Glaciers Online/Jürg Alean; **24** Still Pictures/Brian Lynas; **25** Still Pictures/Mark Lynas; **26** USGS/courtesy Glacier National Park Archives; **27** USGS/Karen Holzer; **28, 29** Geophotos/Tony Waltham; **30a** Greenpeace/Archivio Museo Salesiano; **30b** Greenpeace/Daniel Beltra; **32** PA Photos/AP/Peter Schneider; **33** Greenpeace Schweiz; **35a** Alamy/Dave Pattison; **35b** Alamy/Robert Estall Photo Agency/David Coulson; **36, 37** USGS Earthshots; **38–39** NASA/Goddard Space Flight Center Scientific Visualization Studio; **40, 41** NASA/Goddard Space Flight Center; **42, 43** Science Photo Library/CNES, 1998 Distribution Spot Image; **44** Alamy/Robert Harding Picture Library/Gavin Hellier; **45** Getty Images/AFP/Andrea Merola; **46, 47** Still Pictures/NASA/UNEP; **48** Bilderberg/Klaus D Francke; **49** Corbis/EPA/Holger Hollemann; **50, 51** UNEP; **52** Corbis/Zefa/Svenja-Foto; **53** Corbis/EPA/Martin Gerten; **54, 55** UNEP; **56, 57** Ecoscene/Paul Thompson; **58, 59** Phillip Dustan, FLS, College of Charleston; **60, 61** Still Pictures/UNEP/NASA; **62** Alamy/Juliet Butler; **63** Alamy/Lou Linwei; **64, 65** NASA; **66** Corbis/Reuters/Paulo Whitaker; **68** Getty Images/Hulton Archive; **69** Alamy/Iain Masterton; **70** Corbis/Bettman; **71** Corbis/Michel Setboun; **72** Gulf Images/Image Solutions; **73** Getty Images/AFP; **74** Alamy/Mike Goldwater; **75** Still Pictures/Mark Edwards; **76, 77** Getty Images/Staff/Justin Sullivan; **78** Corbis/Bettmann; **79** Alamy/Charles O Cecil; **80** Corbis/Museum of History and Industry; **81** Corbis/Joel W Rogers; **82, 83** Still Pictures/NASA/UNEP; **84** Corbis/Bettmann; **85** Corbis/Free Agents Ltd; **86, 87** Still Pictures/Mark Edwards; **88a** Corbis/Roger Antrobus; **88b** Alamy/Worldwide Picture Library; **90a** Corbis/Sygma/Bisson Bernard; **90b** Alamy/Matthew Noble; **92, 93** Sund & Bælt Holding/Mezzo Media; **94a & b** EcoScene/Chinch Gryniewicz; **96, 97** Still Pictures/UNEP/NASA; **98** South American Pictures; **99** Alamy/AGB Photo; **100a & b** Still Pictures/Mark Edwards; **102, 103** UNEP/NASA; **104** Getty Images/Hulton Archive/Bert Hardy; **105** Corbis/Cordaiy Photo Library/John Parker; **106** Alamy/Simmons Aerofilms Ltd; **107** Alamy/Andrew Holt; **108, 109** Seoul Metropolitan Government; **110** Corbis/Bettmann; **111** Corbis/Zefa/José Fuste Raga; **112** Getty Images/Time & Life Pictures/Dmitri

Kessel; **113** View/Paul Raftery; **114, 115** UNEP/NASA; **116a** The George A Eslinger Street Lighting Photo Gallery, City of Los Angeles; **116b** Alamy/Chad Ehlers; **118** Eye Ubiquitous/Hutchison; **120** Getty Images/Time Life Pictures/ A Y Owen; **121** Getty Images/Adriel Heisey; **122** Oxford Scientific Photo Library/Mike Powler; **123** Corbis/EPA; **124, 125** NASA/Earth Observatory; **126, 127** USDA Foreign Agricultural Service, Production Estimates and Crop Assessment Division; **128** Rex Features; **129** Alamy/ Oyvind Martinsen; **130, 131** Still Pictures/NASA/UNEP; **132, 133** NASA/GSFC/METI/ERSDAC/JAROS and US/Japan ASTER Science Team; **134, 135** Still Pictures/UNEP/NASA; **136** Still Pictures/Jacques Jangoux; **137** Still Pictures/Luiz C Marigo; **138,139** UNEP/NASA; **140, 141** Still Pictures/UNEP/ NASA; **142–3** Corbis/Gary Braasch; **144** Corbis/Charles E Rotkin; **145** Corbis/Jim Richardson; **146, 147** Still Pictures/ UNEP/NASA; **148, 149** UNEP/NASA; **150** Corbis/Bettmann; **151** Reuters/Mark Baker; **152** City of Vancouver Archives; **153** Robert Harding/Michael Jenner; **154, 155** UNEP/NASA; **156** Reuters/Stringer; **157** Reuters/Gleb Garanich; **158, 159** Corbis/Reuters; **160** Alamy/Picture Contact/Jochem Wijnands; **161** Corbis/Viviane Moos; **162** Panos Pictures/ Jim Holmes; **164** Corbis; **165** Corbis/Yann Arthus-Bertrand; **166, 167** Arctic Images/Ragnar Th Sigurdsson; **168, 169** FLPA; **170, 171** USGS/Earthshots; **172** Corbis/ Jim Sugar; **173** Corbis/Douglas Peebles; **174, 175** Corbis/ Sygma/Patrick Robert; **176–7** Science Photo Library/NOAA/ Robert M Carey; **178, 179** Hedgehog House/Nick Groves; **180** Corbis/Earl & Nazima Kowall; **181** Corbis/Ryan Pyle; **182** Getty Images/Time & Life Pictures/Nat Farbman; **183** Corbis/Bettmann; **184** South American Pictures/Tony Morrison; **185** Corbis/Lloyd Cluff; **186, 187** NASA/Earth Observatory/Space Imaging; **188** PA Photos/PA Archive; **189** PA Photos/PA Archive/John Giles; **190, 191** Corbis/Parks Victoria/Handout/Reuters; **192–5** Space Imaging/CRISP/ National University of Singapore; **196** Corbis/Royalty Free; **197, 198, 199** Corbis/Dallas Morning News/Smiley N Pool; **200, 201** NASA/Visible Earth/Johnson Space Center; **202** Science Photo Library/NASA/Goddard Space Flight Center; **204, 205** Corbis; **206** South American Pictures/Tony Morrison; **207** South American Pictures/Peter Francis; **208** Corbis; **210, 211** Imperial War Museum Collections, London/Crown copyright; **212** Alamy/Popperfoto; **213** Corbis/Michael St Maur; **214** Associated Newspapers; **215** Alamy/Blue Pearl Photographic/Richard Osbourne; **216** Getty Images/Hulton Archive; **217** Corbis/MedioImages; **218** Getty Images/Hulton Archive; **219** Getty Images/ Photographer's Choice/Walter Dieterich; **220** Getty Images/ AFP; **221** Getty Images; **222** Getty Images/Hulton Archive/ D Trakhtenberg; **223** Robert Harding/Sylvain Grandadam; **224** Corbis; **225** Corbis/Caroline Penn; **226** Corbis/Bettmann; **227** Corbis/Sygma/John Van Hasselt; **228, 229** Getty Images/ MPI/Hulton Archive; **230** Getty Images/Hulton Archive; **231** Alamy/David South; **232, 233** TopFoto.co.uk/Topham/ AP; **234** UNEP/Hassan Partow/NASA; **234–5** Still Pictures/ UNEP/NASA; **236a** Alamy/Jenny Matthews; **236b** Getty Images/AFP; **238** Getty Images/Hulton Archive/Alice Schalek; **239a** Corbis/Sygma/Nigel Chandler; **239b** Alamy/Duncan Soar; **240** Corbis/Sygma/Patrick Chauvel; **241** Getty Images/ Spencer Platt; **242, 243** IKONOS satellite image courtesy of GeoEye/INTA. Copyright 2007. All rights reserved; **244** Getty Images/Stone/Joe Pobereskin; **245** Getty Images/AFP/Stan Honda; **246** Still Pictures/David Woodfall; **248** Corbis; **249** Alamy/Travel Ink/Dennis Stone; **250** TopFoto.co.uk/ Topham Picturepoint; **251** Corbis/WildCountry; **252** TopFoto.co.uk/HIP; **253** Getty Images/AFP/Deshakalyan Chowdhury; **254** Corbis/Bettmann; **255** Corbis/Enzo & Paolo Ragazzini; **256** Corbis/Hulton-Deutsch Collection; **257** Giulia Hetherington; **258** Mary Evans Picture Library; **259** Alamy/ Michael Dwyer; **260** photo John Thomson, courtesy William Lindesay; **261** William Lindesay, www.2walls.org; **262** Corbis/ Dean Conger; **263** Corbis/EPA/Wu Hong; **264** PA Photos/AP; **265** PA Photos/AP/Jim McKnight; **266** Corbis/Underwood & Underwood; **267** Alamy/John Peter Photography; **268** Getty Images/Hulton Archive; **269** Corbis/Ric Ergenbright; **270, 271** Eden Project; **272, 273** Phil Sheldon Golf Library/ Richard Castka; **274** Corbis/Brian A Vikander; **275** Corbis/ Reuters/Fayaz Kabli; **276** Getty Images/Roger Viollet/ Pierre Barbier; **277** Corbis/Reuters/Emmanuel Fradin; **278, 279** NASA/GSFC/MITI/ERSDAC/JAROS, and US/Japan ASTER Science Team; **280** Getty Images/Hulton Archive/ Phillip; **281** Corbis/Eye Ubiquitous/David Cumming.

Mitchell Beazley would also like to thank Philippa Bell for her assistance in putting together this book.